We Walk On Jewels

Treasure Hunting in Maine

for Gems and Minerals

JEAN BLAKEMORE

Courier of Maine Books
One Park Drive
Rockland, Maine 04841

Library of Congress Card Catalog Number: 76-40585
International Standard Book Number: 0-913954-26-8

A Courier of Maine Book

Courier of Maine Books are published by
Courier-Gazette, Inc.

Printed in the United States of America

Contents

Preface

For those of you who are interested in the complicated subject of rocks and minerals, but not enough to want to make an extensive study of them, this is what this book will do for you, in a simple, readable way: explain the geology of one of the most minerally endowed areas in the world, the State of Maine; describe the methods, both amateur and professional, by which the minerals may be recovered; identify the minerals and metals as simply as possible aided by many illustrations in color; locate hundreds of mines and outline the many uses to which minerals may be put by you.

For you rock hounds, ideally with your families, this will be a complete field guide to about eight hundred mines in all parts of Maine. The instructions for creating important mineral collections and for cutting gems and setting them into jewelry will fire you with ambition for using the treasures you bring home.

Thomas Huxley wrote, "To a person uninstructed in natural history sciences, his country or seaside stroll is a walk through a gallery filled with wonderful works of art, nine-tenths of which have their faces turned to the wall."

It is my hope that this book, whether you stroll in Maine or anywhere else in the world, will turn many of the pictures face to you for your enjoyment and enrichment.

My grateful thanks to four of Maine's outstanding mineralogists: Robert Doyle, State of Maine Geologist and his assistant, Walter Anderson — their help and advice has been invaluable; Priscilla Bryant, owner of the Winthrop Mineral Shop — she made selections from her wealth of Maine minerals which were used in the color photographs; James Mann, Jr., of East Bethel — he provided most of the specimens and cut gemstones used in the color photographs. I also wish to thank my daughter, Jane Blakemore Sowton, whose amethyst is photographed in color. The lapidary and jewelry-making photographs are of my old friend, Charles Cole, fine silversmith and jeweler. The color photographs are by Paul Fournier of Hallowell. The diagrams of the color photographs are by Anne DeMaria.

Alphabetical List of Maine Minerals with Some Favorable Places for Collecting

Acmite
Phippsburg

Actinolite
Lime quarries
Rockland
Thomaston

Albite
Buckfield

Allanite
Brunswick

Alunogen
Newry

Amazonite
Southwest Harbor

Amblygonite
Paris

Amethyst
West side
Pleasant Mt.,
Denmark

Amphibole
Thomaston

Analcite
Greenwood

Andalucite
Road cut
Carabasset
Bay Point

Andecine
Denmark

Ancrthite
Sanford

Apatite
Auburn
Buckfield
Greenwood
Noyes Mt.

Apophyllite
Perry

Aragonite
Thomaston

Argentite
Sullivan

Arsenic
Greenwood

Arsenopyrite
Winslow tin mine
Verona Island

Asbestos
Little Spencer Stream

Augite
Phippsburg
Cape Rosier

Autunite
Topsham

Axinite
Phippsburg

Azurite
Blue Hill

Barite
Brunswick
Deer Isle

Bauxite
In clays around
Passamaquoddy
and Penobscot bays

Bertrandite
Stoneham

Beryl
Buckfield
Albany
Brunswick

Beryllium
Oxford
Androscoggin
Cumberland
Sagadahoc counties

Beryllonite
Stoneham

Biotite
Topsham
N. Edgecomb

Bismuth
Sullivan

Bismuthite
Brunswick

Bornite
Blue Hill

Brookeite
Paris

Calcite
Lime quarries
Thomaston

Cancrinite
Litchfield

Cassiterite
Winslow
Greenwood
Black Mt.
Rumford

Cerarggyrite
Sullivan

Chabazite
Phippsburg

Chalcedony
Jasper Beach

Chalcocite
Blue Hill

Chalcopyrite
Blue Hill

Chiastolite
Standish

Childrenite
Newry

Chlorite
Blue Hill
Union

Cholentite
Red Ledge

Chromite
Deer Isle

Chrysoberyl
Hartford

Chrysocolla
Beddington

Chrysolite
Boothbay

Chrysotile
Little Spencer Stream

Cimolite
Norway

Clay, marine
everywhere, especially
the coast, Columbia
Falls, Machias,
Windsor, N. Waldoboro,
Newington

Columbite
Newry

Cookeite
Paris

Copper
Hancock County
Union
Woolwich

Cordierite
Gardiner

Corundum
Greenwood

Cuprite
Blue Hill

Cyanite
Gorham

Dahllite
Buckfield

Denderite
Bowdoinham

Diopside
Cornish
Phippsburg

Dioptase
Beddington

Dolomite
Warren
Union

Eosphorite
Newry

Epidote
Thunder Hole
Mt. Desert Island
Raymond

Fairfieldite
Buckfield

Feldspar
Everywhere

Fluorite
Mt. Waldo
Frankfort
Stoneham

Fossils
Moose River

Francolite
Newry

Gahnite
Topsham

Galena
Blue Hill
Cape Rosier
Lubec, Union

Garnet
Hebron, Minot,
Topsham, Cornish
Phippsburg

Gold
Swift River, Byron,
Sedgewick and more
than fifty other
locations.

Graftonite
Greenwood

Granite
Everywhere

Graphite
Crocker Hill, Paris,
Andover, Lewiston

Gravel
Everywhere

Gummite
Auburn

Gypsum
Messalonskee Gorge
Oakland, Saco

Halloysite
Paris

Hamlinite
Greenwood

Hatchettolite
Newry

Hematite
Castle Hill, Mapleton

Herderite
Buckfield

Hornblende
Bayville, Greene,
Jonesport

Hydronephelite
Litchfield

Hypersthene
Wayne

Idocrase
Cornish

Ilmenite
Newry

Iron
Mt. Katahdin
Union

Iron (meteoric)
Castine

Kaolinite
Rockport

Kaylinite
Concord

Kunzite
Andover

Kyanite
Standish, Windham

Labradorite
Appleton

Landesite
Poland

Laumonite
Phippsburg

Lead
Aroostook County
Cape Elizabeth
Hancock County

Lepidmelane
Litchfield

Lepidolite
Black Mt., Rumford
Auburn

Limonite
Katahdin
Iron Works
Union

Limestone
Rockland
Thomaston

Linneasite
Union

Lithiophilite
Hebron

Lollingite
Hebron

Magnesium
Aroostook County

Magnetite
Topsham

Malachite
Blue Hill
Woolwich

Manganese
Aroostook County

Manganite
Newry

Mangetite
Brunswick

Marcasite
Jewell's Island
West Paris
Winslow

Marialite
Raymond

Melanterite
Concord

Mica
Everywhere

Microcline
Mt. Desert

Microlite
Newry

Molybdenite
Blue Hill,
Cooper, Alna
Township 10

Molybdite
Cooper

Monazite
Brunswick

Montmorillonite
Greenwood
Buckfield

Muscovite
Newry, Rumford

Natrolite
Windham

Nephelite
Litchfield

Obsidian
Castine

Oligoclase
Lewiston

Olivine
St. George
Boothbay

Opal
Warren

Orthoclase
Pownal,
Trott Cove,
Woolwich

Ottrelite
S. Portland

Pearls
Salt
Boothbay
and other
Coastal
sweet
Dresden
N. Farmington
Oxford County

Peat
Centerville
DeBlois
Franklin

Petalite
Greenwood

Phlogopite
Hebron

Phochlorite
Buckfield

Phonocite
Stoneham

Pitchblende
Newry

Plagioclase
Trott Cove,
Woolwich

Platinum
Hermon

Pollucite
Hebron
Newry
BB#7

Prehnite
Portland

Psilomelane
Buckfield
Mapleton

Purpurite
Newry

Pyrargyrite
Franklin

Pyrite
Buckfield
New Limerick

Pyrochlore
Buckfield

Pyrolucite
Mapleton
Rumford

Pyromorphite
Lubec

Pyroxene
Raymond
Union

Pyrrhotite
Katahdin Iron
Works,
Union

Quartz
Everywhere
Newry, Auburn,
Albany, Paris,
Buckfield, Brunswick
Greenwood

Reddingite
Buckfield

Rhodochrosite
Buckfield

Rhodonite
Blue Hill

Rutile
Litchfield, Warren

Samarskite
Topsham

Scheelite
Cooper, Byron

Serpentine
Deer Isle, N. Haven,
Spencer

Siderite
Newry

Sillimanite
Paris, Warren

Silver
Pembroke
Sargentville
Sedgewick, Sullivan

Sodalite
Litchfield

Specularite
Cadillac Mt.,
Mt. Desert

Sphalerite
Dexter, Cape Rosier

Spinel
Greenwood

Spodumene
Rumford, Warren

Staurolite
Windham, Winthrop

Stephanite
Gouldsboro

Stewartite
Newry

Stibnite
Hallowell
New Limerock

Stilbite
Hallowell

Stromeyerite
Sullivan

Sulphur
Katahdin

Talc
Cape Rosier
Sidney, Warren

Tannanite
Blue Hill

Titanite
Topsham

Topaz
Topsham

Torbenite
Newry

Tourmaline
Newry, Black Mt.,
Greenwood, Hebron,
Paris, Auburn

Triphylite
Buckfield

Triplite
Stoneham

Trmolite
Lime quarries
Rockland
Thomaston

Uraninite
Newry

Uranophane
Auburn

Uranopilite
Newry

Vesuvianite
Sanford

Vivianite
Madawaska

Wernerite
Raymond

Wolframite
Bowdoinham

Yttrocerite
Paris, Newry

Zinc
Castine

Zircon
Litchfield

1. Maine's Buried Treasures

Do we ever outgrow the magical lure of treasure hunting!

When I was a little girl my friends and I dreamed of riches. We pried garnet crystals out of Boothbay Harbor's rocky rim, dove for mussels for their pearls and marvelled at the veins of blood red quartz and the smooth faces of big black tourmaline crystals.

After Labor Day every Fall the State of Maine Express bore us to New York. Each clutched two small vials, one with pearls and one with garnets. We intended to take them to Tiffany's, the noted jewelry store. Somehow none of the bottles ever reached there. Now we smile to see our children following the same plan.

In those long ago days I used to gaze out over Linekin Bay. Her perfectly placed islands tufted up like Gothic cathedrals and English crumpets. Beyond the outer islands the deeper blue of the open sea stretched to the horizon. I used to say to myself, *"All this beauty! And we walk on jewels, too!"*

A spark was struck, never since extinguished, when an old man told me that every known gem, mineral and metal has been found in the State of Maine.

Exaggeration? Yes, but only slight.

"Maine," said a former Maine geologist, Professor James M. Trefethen, "has a greater variety of minerals than any other state in the nation. And Oxford County has a greater variety than any area of its size in the world."

"In southwestern Maine," wrote the late Stanley I. Perham, mineralogist of West Paris, Maine, "as in few areas of similar size in the world, are found some two hundred varieties of minerals and gemstones. Collectors come here from everywhere to explore the beautiful Maine hills for what they know lies hidden there."

Newry, Black Mountain and Whitehall in Oxford County are so favored with the miracle minerals of the future they make uranium old hat, though of course Maine has uranium, too!

John Rand, former State Geologist, told me that if large mining companies with research laboratories should operate these areas and develop uses for these wonder materials as they are recovered, Maine may be aiding space programs as yet undreamed of. Isn't that an imagination stretcher!

May take some doing, though. One such company once spent a year and a half and two hundred thousand dollars for a preliminary survey of the beryllium prospects at Newry. Their conclusion was that this rare metal, though present in huge quantities, was too at random with too many blank areas to be economically feasible to get out.

John Rand refers to these rarities as "oddballs."

Our present State of Maine Geologist, Robert Doyle, speaks of them as "farouts."

Among these are the space minerals:

the cesium (also spelled caesium) group: beryl and bertrandite and the clay into which they sometimes disintegrate, montmorillonite; pollucite (15% cesium) and yttrocerite.

the lithium (the lightest known metal) group: autunite (a uranium bearer); lepidolite; torbenite and niobium (columbian with a bit of tantalite). Tantalite will make a geiger counter needle shoot off the dial and its light flash faster than you can count.

Very far out are: Nepheline syenite; managana-apatite; allanite and mangana-tantalite.

It would take all the superlatives of a P. T. Barnum to describe our Maine minerals. Only all this is true. So, here goes —

LARGEST

In 1952 when I wrote the first edition of this book, I described the world's largest beryl crystal found in 1949 in Bumpus mine at Albany, Oxford County, Maine. I gave its dimensions as eighteen feet long, forty-two inches in diameter and twenty-six tons in weight. By the way, you can see this single crystal of aquamarine if you enter the 79th Street door of The New York Museum of Natural History. The crystal broke in two as it was being recovered and its two parts flank the staircase facing that door.

My description led to a bit of teasing. John Allen May, in his first writeup about me in The Christian Science Monitor entitled, "Maine Doesn't Leave a Stone Unturned," wrote, "Did you know that the world's largest beryl crystal — (we will allow a short pause here for Texans to get adjusted to the fact)."

"Did you know, too, that this blue-green beryl crystal was 18 ft. long, 42 in. in diameter and weighed 26 tons? (Let us allow another pause for every man jack of us to get adjusted.)"

In describing, "with wide-eyed interest closely akin to horror," the equipment I recommended for the hunt he adds, "—also, perhaps, a large truck and a crane in case he comes across a fair-sized Maine beryl crystal."

Then, when John Rand was State Geologist, he remarked to me, "It's a wonder to me where people get their misinformation! Yesterday I asked a man in Rumford who had just written a magazine article on the Bumpus crystal where he had gotten his specifications."

I blushed scarlet. "I gave those same specifications in my book."

"I know," he said dryly, "because he answered, 'from Jean Blakemore's book'."

But I have been vindicated because I have learned since that in 1950 Bumpus mine beat its own world's record for the third time.

Let Dana Douglas, in charge at the time of the spine tingling find, tell about the event.

"When the smoke of the dynamite charge cleared I nearly fainted. Imbedded in a cliff of rose quartz was an unbelievably large single beryl crystal.

"It was twenty-seven feet, nine inches tall and weighed twenty-seven tons.

"It looked for all the world like a giant ice cream cone, upside down, graduating from eleven inches at its rounded top to six and a half feet at the bottom. It was all shades of blue and blue-green (aquamarine). Incorporated within the crystal and sprouting around its base were thousands of small crystals (phantoms we call such).

"To complete the amazing picture, a scoop of 'vanilla ice cream' was there, too, right in its proper position. At the bottom of the aquamarine cone was a curved line of demarcation below which, scoop shaped, were six tons of heliodore, the rare golden beryl."

Did Dana preserve this marvel for future generations to admire in The Smithsonian Institute? He did not!

He ordered it broken into chunks. Fifty-two thousand pounds of it were loaded onto box cars and shipped off to The United States Government. It is in great demand, and short supply, for the beryllium derived from it. This world rare "space" metal will shield against uranium better than anything else on earth, an eighth of an inch compared to the three feet required of lead.

But, never think that Maine has but one string to her "largest" bow!

Beryl, quartz, chrysoberyl and tourmaline — all these gem families have established records for Maine.

Beryl

The LaChance quarry in Brunswick produced one of the largest gem beryls ever found.

The largest gem aquamarine in North America came from Lord Hill in Oxford County. This mountain yielded the largest gem crystals of beryllonite.

Quartz

A double terminated quartz crystal twenty-two inches thick, thirty-six inches tall, weighing eight hundred fifty pounds was found in the Paul Bennett quarry in Buckfield. This was the largest on record in New England.

Harvard Museum has on display the largest known mass of quartz crystals, found by the late Ike Skillen of Freeport, Maine.

The aforementioned Bumpus mine has the largest concentration of rose quartz in the world.

At Red Ledge in Oxford County a pocket was opened containing the largest known quantity of the very rare rose quartz crystals.

Chrysoberyl

Hooper's Ledge in Oxford County produced a three-pound mass of green chrysoberyl. Another world's record.

Tourmaline

Mt. Mica on Paris Hill, where tourmaline was first found in Maine, has produced the largest green tourmalines. In 1898 the late Lorin B. Merrill found here a flawless blue-green tourmaline nodule weighing four hundred and eleven carats. Even in those days it was valued at one hundred thousand dollars. The first finding of tourmalines was a stranger than fiction story I am saving until we get to this mine in the fifth chapter.

Another exciting tale involves the opening up of the largest pocket of gem emerald green tourmaline crystals in history.

In 1955 newspapers, magazines and radio, even in England, were shouting for weeks about Stan Perham's lucky blast.

Early one morning in his BB No. 7 mine near Norway in Oxford County, Stan put in a light charge of dynamite where rounded rolls of pale blue cleavelandite, often a clue, led him to suspect there might be a pocket. He was stunned when the six by two foot cavity revealed a fortune in the costly gem, green tourmaline. He said it took three days before he recovered enough to realize what he had found.

There were six thousand carats of lovely terminated crystals, the largest weighing one hundred thirty-seven carats.

He spent the whole day picking out of the soft white clay, cookeite, all these hundreds of crystals which had originally grown out of the pocket walls later to be broken away by glacial pressure. He said it was just like picking raisins out of a plum pudding.

Nightfall found him still at it and resigned to working through the night rather than leave any of that fortune exposed in the pocket. Luckily his wife and daughter came looking for him and stayed to help. By sunup the three had filled pails with two hundred thousand dollars worth of gems.

The largest and richest tourmaline find in the world happened in Oxford County in 1972. The story begins in 1937 when Dick Nevel, renowned for his uncanny nose for minerals, was mining for pollucite at Newry Old Mine, up the mountain from the newer one. He was so sure that he was about to open up a pocket of tourmaline that he painted his name on the rock face before putting in a charge of dynamite. He was killed by the blast, and the debris in the blast area went undisturbed for 36 years.

Then, in 1972, Dale Swett was advised by his doctor to give up his desk job and do some relaxing work outdoors. Dale is a rock hound who lives near Newry Mountain; so there he began to explore. One day, happily relaxed, he scratched away an overburden of loose debris.

From the moment he found Dick Nevel's name and uncovered the pocket of tourmaline Dick's blast had opened, he wasn't to know a relaxed second for three years. For Dale had exposed the first of 14 pockets crammed with two million dollars worth of Maine's official gemstone, tourmaline.

True to its name, "the harlequin of stones," the tourmaline had many colors — green, pink, ruby red (rubellite), blue (indicolite), yellow, and a new color, grape jelly. They ranged from exquisite tiny crystals to spectacular giants.

A corporation of experts was formed, The Plumbago Mining Company, with Dean McCrillis, president, was formed to carry on the work. Priscilla Bryant, in her Winthrop mineral shop, handles the wholesale and retail sales of the fabulous material.

To visit the locale, you should first check with Dean McCrillis in Roxbury.

STILL MORE 'LARGEST'

In 1933, in Topsham, Sagadahoc County, a cavern of jewels was opened, a pocket twelve feet long by ten feet wide by six feet deep. From the top hung scores of quartz crystals. Imbedded deeper in the top lining brownish white herderite (a beryllium ore) crystals were interspersed, among them the world's largest. The pocket contained many other interesting specimens, among them the radioactive mineral torbernite, a phosphate of copper and uranium.

And on the floor of this treasure cave was a considerable quantity of topaz, some blue and some white. One, weighing ten pounds is said to be the largest topaz ever found in this country. Did you know topaz comes in other colors than the brownish yellow we associate with it? The blue is almost indistinguishable from aquamarine. The National Museum in Washington has a 43¾-carat blue topaz gem found by John Seaver Towne in a tourmaline pocket at Mt. Apatite, Auburn, Androscoggin County, Maine.

Lithium

In 1957, lithium was found in Warren, Knox County. This may be the largest bed in the East of that lightest of metals, which is used to trigger the atom bomb. More recently lithium has begun to be used as a medicine.

Cesium

The T. C. Mining Company of West Paris owns twenty mines in Oxford County estimated to contain three million tons of cesium, ninety-five percent of what exists in the free world. In 1954 at Noble's Corner, Norway, Maine, this company recovered sixteen thousand grams of the ore, the largest concentration in existence.

In 1959 Dr. Willis F. Thompson of Portland, former consulting engineer for the atomic energy commission, found in West Paris the world's largest deposit of virtually pure ore (42% and up cesium).

This rare element has many important strategic uses. It is used in photoelectric cell construction, for scanning apparatus, infrared signalling devices, electronics and nuclear research. Because it burns with terrific intensity in a vacuum it is used in solid fuel for space rockets. The Los Alamos New Mexico Scientific Laboratory has developed a method of converting nuclear energy into electric power through the use of a device, thermocouple, employing ionized cesium gas.

Catherine Hill on the shores of Tunk Pond, Hancock County, has the world's largest deposit of molybdenum, "white gold," which produces molybdenum steel, the fine alloy used in airplane construction.

MAINE FIRSTS

The first gold discovered in the United States was not in California, not in the Yukon. No, indeed! It was discovered at the Swift River, Byron, Maine, in the crop of a chicken. When the farmer discovered those shiny yellow flakes in the crop of his Sunday dinner he remembered he had dug the hen house scratch from the gravelly banks of the Swift River. It was in the early 1800s that the assay report came back "pure gold." And from that day to this hundreds of us, those of us who know the technique of panning, have been showing color from the river. Later in this book I'll explain how to do it.

The first necklace of U. S. gems was of Maine tourmalines. It was worn in Washington society by the wife of Maine's Hannibal Hamlin, Vice President under Lincoln.

In 1861 lead and zinc were first mined in this country in Hancock County, Maine.

The Portland Slate Company in Monson, Maine, made the first slate sink, which is still in use, for The World's Fair in 1898. The slate butter boxes contrived to keep cool and fresh the butter used by the miners' families were the first insulated boxes.

The first book on minerals ever published in America was "An Elementary Treatise on Mineralogy and Geology" by Professor Cleaveland of Bowdoin College published in 1816.

ONLY

Only in Maine has beryllonite been found.

Monson Maine, is world famed as the only locality in the U. S. which produces slate which satisfies the exacting specifications required for electrical purposes. Slate from Monson was used in the construction of New York's Empire State Building.

Winslow, Kennebec County, Maine, is the only source of cassiterite, tin ore, in the U. S.

THE FINEST IN THE LAND

C. W. Clark of Farmington, found that the entire bed of Lake Umbagog northwest of Rumford is all asbestos. Johns Manville said the report they received on it was the finest they had ever had on asbestos.

In 1835 the finest emerald ever discovered in North America was found in Topsham, Sagadahoc County. It was a one hundred and thirty-three carat stone so pure and exquisitely beautiful it ranks as one of the notable gems of the world.

FAMED THE WORLD OVER

Once it was thought that the only source of pollucite, the cesium bearing mineral, was Elba, the island of Napoleon's exile, and there only in ounces. Then, the late Perien Dudley made the first discovery of it in America at Mt. Rubellite in Hebron, Oxford County. He found one mass six feet long, twelve to fourteen inches thick weighing fifty pounds. Later pockets yielded up to twelve hundred pounds. Since then it keeps cropping up in various parts of Oxford County. Once when I climbed up on a pile of white rock to eat my picnic at Stan Perham's BB No. 7 mine, Stan said, "You are sitting on five hundred pounds of pollucite worth six hundred thousand dollars."

Royal purple amethysts of the finest water were found on Pleasant Mountain in Denmark, Oxford County by the late George Howe. Recently this same mountain and also Deer Hill produced most valuable and beautiful amethyst crystals.

The LaChance Quarry in Brunswick boasts finds of top quality gem beryl and the finest herderite crystals, blue topaz and sheet mica.

Pit Pulsifer in Auburn, Androscoggin County, is noted, too, for finest aquamarine, tourmaline, royal purple apatite, fancy gem smoky quartz and the rare deep pink cesium beryl gem morganite.

There is no museum or private collection of any importance in the world which does not have a specimen of vesuvianite from the famous deposit in Sanford, Cumberland County, Maine. The ancient Romans first found it in the crater of Mt. Vesuvius, thus it came by its name. I, too, found a specimen there which I brought home to Maine to place next to my choicest specimen of Sanford vesuvianite.

VALUABLE

Since 1820 some hundred million dollars worth of Maine minerals have been sold. Less than ten million of this has been gemstock.

To date over two million tons of potash feldspar have been shipped from the pegmatites of Maine for use in ceramic products, abrasives, cleaning compounds and electrical insulators. Unfortunately substitutes have been found and the demand for feldspar has been dropping. Many of the once flourishing mines and processing plants have closed down.

In other directions, though, world scarcity of strategic minerals and metals and modern development of more efficient mining methods are stepping up Maine's mineral production.

Mr. Thomas of The Penobscot Mining Company, an international group formerly engaged in developing a two thousand acre tract at Perry, Ayer, Lubec and Cape Rosier in Hancock County, declared it to be one of the best exploration possibilities he had ever seen. The area contains fabulous deposits of zinc, copper, lead, manganese, some uranium and some silver.

Over one and a half million tons of peat are present in Maine. This is in demand for fuel and peat moss for nurserymen.

Sand and gravel deposited at the end of the last glacial period — ten to twenty thousand years ago — is the largest mineral industry in the U. S. in terms of tonnage. Maine produces over seven million tons a year.

MAINE'S NO. 1 RESOURCE

And what do you suppose it is? WATER! The water power of her rivers. When the projected dam at Passamaquoddy Bay has harnessed the twenty foot tide it will be unique in the world as a source of cheap tidal power. Too, Maine is fourth in the U. S. in the production of mineral waters.

2. The Rocks Beneath Us

Don't submit to the tyranny of the "should" by trying to digest these facts all at once. Skip to your heart's content. The information will be right here when you are hungry enough for it. Let the minerals themselves teach you — the look and the feel of them.

Their fascination will grow on you and lead you on until you will be surprised to find you have absorbed through your skin a university course in minerology. Well — almost!

THE GEOLOGY OF MAINE

The bedrock of Maine was formed during Pre-Cambrian and Paleozoic Eras. The Pre-Cambrian, mostly in the western and southern parts, consists of sandstone, shale and limestone. Where they were compacted to a crystalline rock by magmatic intrusion pegmatites were formed, in which are found valuable mineral deposits.

In central and northern parts, are represented most of the periods of the Paleozoic Era. These rocks had been less altered by metamorphism and contain fossils.

The east branch of the Penobscot River shows beds of quartzite and slate of Cambrian origin and Cambrian fossils.

In the central and northern parts, along the coast at Eastport and in Aroostook are the Silurian sedimentary rocks. These are shales, silts and impure limestones. Silurian fossils are found here.

There is a Devonian belt of sandstone, the Moose River formation, from west of Moosehead Lake to Aroostook. This is the most fossiliferous formation in the state.

During the Devonian Era most of Maine's granites came into position by the intrusion of great bodies of granite called "batholiths."

During the same period volcanoes left beds of lava and ash, deposits of which are seen along the coast from Penobscot Bay to Eastport. Inland, Mount Kineo is the most famous. Others are East Kennebago, Big and Little Spencers, the Coburn Hills in northeast Aroostook, Haystack and the Quoggy Joe Mountains near Presque Isle.

The Paleozoic rock, quartzite slate, borders the coast from Kittery to Casco Bay.

The Devonian Era ended in the Appalachian Revolution, at which time mountains once more rose across New England.

Eras of erosion followed until the Ice Age covered Maine with a continental glacier. This smoothed off the hills and left large deposits of gravel, sand and clay which, damming the pre-glacial valleys, formed lakes and waterfalls.

The weight of the glacier depressed New England. The ocean rose as water from the melting ice poured into it, flooding the coastal lowlands up into the larger valleys. The receding flood left a layer of sand over the clays along the coast. The deserts of Maine at Freeport and Leeds originated this way.

THE THREE TYPES OF ROCK

All rocks fall into one of three catagories: igneous, sedimentary and metamorphic.

Igneous

These are rocks which have hardened from a molten mass called "magma," upthrust from the depths below. They are the earth's original crust.

When, like the lava erupted from a volcano, they have flowed out on the surface, they are "extrusive."

In the case where the rocky melt does not attain the surface it is "intrusive." These rocks exposed when the overburden above has eroded, are coarser grained because they cooled more slowly at their greater depths.

Granite is the commonest of this coarse grained group.

Pegmatite is the name for giant granite so coarse that the various constituents can be recognized easily. A single grain may be several feet across.

It is in the pegmatites of Maine, especially in pockets, that we are most likely to find gemstock. A pegmatite belt extends from Boothbay in Lincoln County to Rangeley in Oxford County in the southwestern part of Maine.

Its pockets have yielded some of the richest gem finds in New England.

PEGMATITES

These coarser offspring of the light colored granites occur as pod shaped or sheet like bodies from a few inches up to half a mile in length.

They are chiefly composed of quartz and orthoclase feldspar with a few grains of a dark mineral, which may be mica (biotite — the black), amphibole or pyroxene.

Sometimes the simple pegmatites were later attacked by solutions containing other more volatile and rarer minerals. Our Maine gem pegmatites are among these complex formations.

Geodes, or pockets, from thumbnail to room sized are sometimes present in the ledge.

Since pegmatites are veins (small) or dikes (large) which have cut across the granite there will either be a definite line or sometimes a merge where the dike meets the granite. This is referred to as a "contact."

So, when light colored rock — GRANITE — gives way to much coarser rock containing random blobs or crystals or masses of light colored glassy quartz, blocky feldspar recognizable by its flat pearly surfaces and sheets of mica (isinglass) built up leaf upon leaf like pages in a book — it is — PEGMATITE.

There are two types of pegmatites depending on the type of feldspar which predominates.

Potash Feldspar Rich

Colored quartz, tourmaline and beryl are to be found in this type.

If it is a pockety area they are often found as crystals growing out of the walls of the cavities in the ledge.

These pockety areas occur in Maine in the Appalachian Mineral Sheet, a zone twenty-five miles long and eight to nine feet wide extending in a northwesterly direction from Auburn in Androscoggin County to Greenwood in Oxford. Here the pegmatites are pod-shaped like a flattened football or peapod, ranging up to 1500 feet long and several hundred feet thick. The pods are generally enclosed in mica schist country rock.

A smaller, non-pockety area of sheet-like pegmatites lie like skull caps on Newry Hill, Whitecap and Black Mountain in Oxford County. Here the gem minerals are imbedded in the solid pegmatite. These pegmatites are generally quite flat-lying, ranging up to more than a mile in diameter and several hundred feet thick.

Soda Feldspar Rich

This type is fluorine rich. In this matrix are to be found amethyst, topaz, certain types of apatite, fluorite, herderite and hamlinite. In Maine this area comprises only a few miles in and near Topsham in Sagadahoc County and in the vicinity of Stowe in Oxford County.

Sedimentary

Rocks of any sort exposed to the various elements break down to soil. These deposits may have been water-laid, wind-laid or ice-laid. Some become cemented together into rocks. Wherever they pile up, as in valleys, lake beds or sea floors, there occurs a horizontal layering called stratification.

Metamorphic

These are any rocks which have been recrystalized by heat or pressure, usually by later intrusions of molten magma.

3. How To Know Them When You See Them

IDENTIFICATION OF THE MINERALS

Of the over two hundred mineral families reported in Maine I am going to explain how to identify easily only the six minerals and the six metal ores you will be most likely to find.

These minerals are quartz, feldspar, mica, tourmaline, beryl and garnet. The metals are gold, silver, iron, lead, zinc and copper.

Their identification depends on one or more, seldom all, of the following characteristics.

1) Luster

Whether it appears metallic or glassy or greasy, or whatever, is one of the best giveways.

2) Color

There are two types:

a) the inherent color, characteristic of the mineral.

b) exotic color, due to presence of impurities or fracturing. For example, quartz is inherently colorless. But it has many family members which have been given names according to their color. Amethyst is the purple colored quartz, citrine is the yellow and so on. This is true of the other families, especially those in which gems occur. Diamonds are of many colors. Sapphires and rubies are members of the same family, corundum. Emeralds and aquamarines are both beryls. And as for tourmalines! You have only to name the color and you have it in that harlequin of stones, the tourmaline family.

3) Streak

This is the color of the powdered material. To determine this you mark a specimen on a white surface such as the unglazed back of a tile. Streak plates are purchaseable.

4) Cleavage

This is the ability of the mineral to split more readily in certain directions than in others. Determine this by holding up to the light a freshly broken, unweathered specimen and rotating it in all directions. The cleavage surfaces will reflect the light brilliantly like little mirrors. Note the perfection of cleavage and the angular direction between two or more cleavage directions.

Mica has the most perfect cleavage known in one direction and in that one direction only can be scaled down to infinitely thin flakes. Feldspar shows smooth pearly surfaces when broken along its cleavage planes. Others, such as quartz, have no cleavage directions at all.

5) Fracture

This is the way the mineral breaks. For instance, quartz shows conchoidal fracture, clamshell-like scoops. Horneblend tends to have a splintery fracture.

6) Hardness

This is the ability of FRESH material to resist scratching. Briefly, it if can be scratched by a knife it is obviously softer than the knife. If it cannot it is either equal to or harder than the knife. If it scratches a knife it is the harder.

The Mohs scale classifies the hardness of minerals from 1 (softest) to 10 hardest)

1) talc	6) feldspar (orthoclase)
2) gypsum	7) quartz
3) calcite	8) topaz beryl
4) fluorite	9) corundum
5) apatite	10) diamond

You can buy a set of hardness pencils. In the back of the book I will give a list of mineral stores where you can buy what equipment you may want and get further help and information.

Lacking hardness pencils it is useful to know that a copper coin is 3, window glass is 5, a key is 5 and a knife is 5 to 6½.

7) Structure

Structure may be granular, fibrous, botryoidal (rounded masses like viscera or bunches of grapes grown together) and crystalline (structure not apparent).

8) Specific Gravity

This is its weight compared with the weight of an equal volume of water. For instance, a cubic inch of quartz weighs 2.65 times as much as a cubic inch of water.

9) Magnetism

A few are attracted by a magnet. Magnetite is the most common of these.

10) Association

Which mineral pals around with which is significant. As an example, olivine seldom associates with quartz. So, if olivine seems to look like some other mineral the absence of quartz may be the identifying clue.

Usually not all of the above clues to identification will be necessary. A couple are often enough.

Our State of Maine Geologist, Robert Doyle, sums it up:

First look for luster. Metallic is one extreme with glassy at the other end of the scale with dull, silky, greasy and pearly in between.

Note whether the color is light or dark.

Judge the weight just by hefting.

Determine the hardness by scratching.

Notice its general appearance where it breaks along shiny crystal faces and the look of its broken surfaces.

If you find a crystal, note its form.

If all of the above leave you still in doubt, get the color of its streak as described above.

Sounds easy, doesn't it?

11) Crystal Form

Internal atomic arrangement in definite geometric patterns is sometimes, but not always, expressed as crystal form.

The various forms fall into six crystal systems according to the position of the axes which are imagined as the skeleton on which the crystal is built. For instance, quartz, when it crystallizes, occurs as a six-sided prism often terminated by a six-sided pyramid.

Often during crystallization, conditions may have been unfavorable to the growth of regular crystals, so you will more usually find the minerals in massive, non-crystallized form.

The following pages show you easy identification of the families of minerals and metals you are most likely to meet in the State of Maine.

Mineral

Quartz	Garnet
Tourmaline	Feldspar
Beryl	Mica

Metals

Pyrrhotite	Chalcopyrite
Galena	Silver
Sphalerite	Gold

The Quartz Family is Easy to Know

* Crystals six-sided with pyramids
* Light colored and glassy
* Breaks with shell-like scoop
* Transparent, more or less

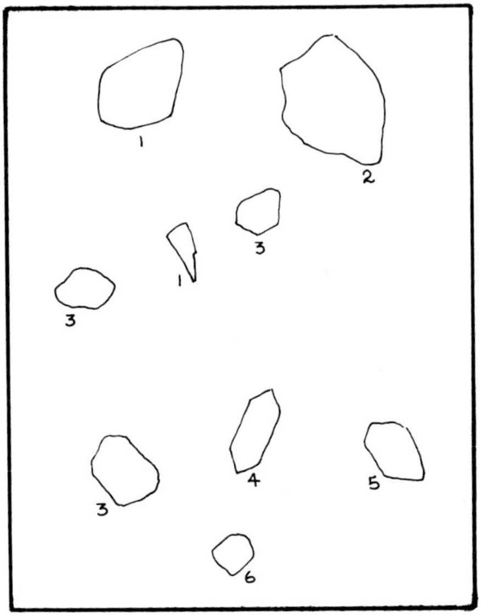

1. Smoky Quartz
2. Amethyst
3. Rose Quartz
4. Rock Crystal
5. Citrine
6. Amethyst Crystal

Aventurine (green) not shown

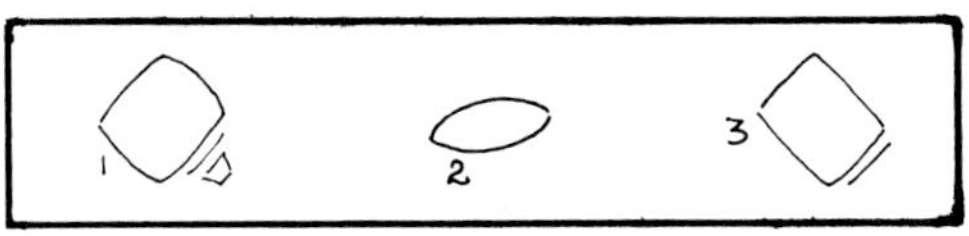

Gemstones

1. Amethyst (facetted)
2. Citrine (facetted)
3. Smoky Quartz (facetted)

The Tourmaline Family is Easy to Know

* Long, striated, pencil-like crystals
* Glassy
* Transparent, more or less
* More often than not found in crystal form

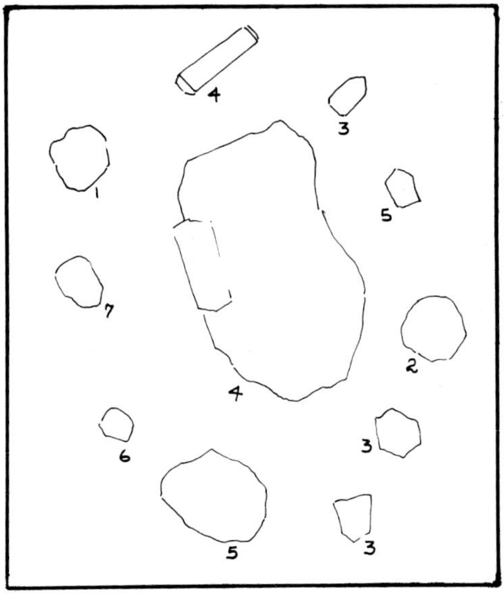

1. Rubellite Crystal
2. Green
3. Watermelon
4. Idicolite
5. Black (Schorl)
6. Rubellite Crystal
7. Grape Jelly

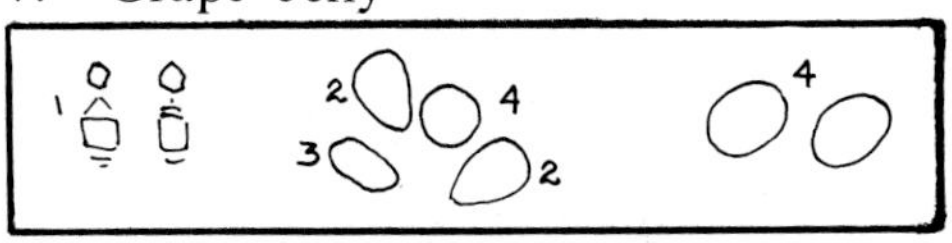

Gemstones

1. Emerald Green (emerald cut)
2. Watermelon (cut freeform)
3. Green (cut freeform)
4. Indicolite (cabachon cut)

The Beryl Family is Easy to Know

* Six-sided crystals
* Transparent to translucent
* Glassy
* Very hard
* Color is its distinguishing feature (from pale green to the greenish-blue of aquamarine)

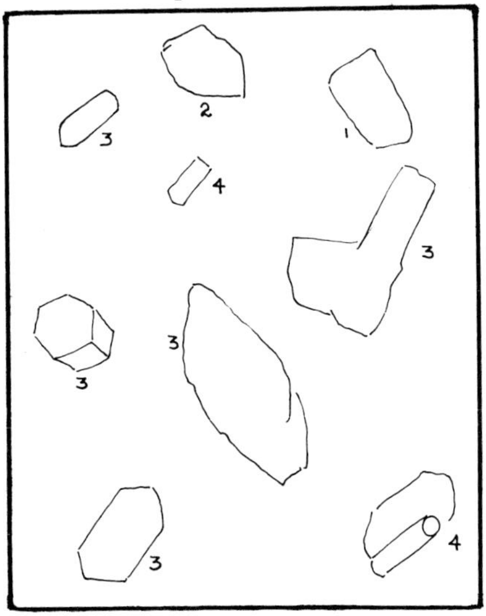

1. Emerald
2. Sugar White
3. Aquamarine
4. Golden

Gemstones

1. 3-piece set uncut aquamarines in the pin. BUT — beware! Look alikes are not always alike. The matching stones in the ring and earrings are from a totally different family. They are blue topaz.
2. Heliodore (facetted)
3. Aquamarine (cut freeform)
4. Morganite (facetted freeform)

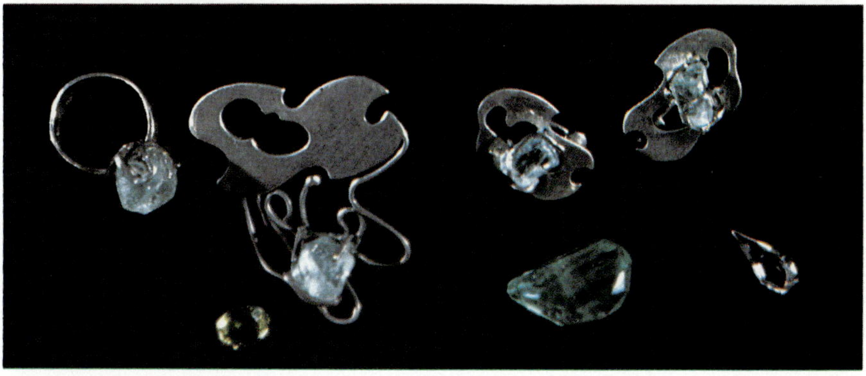

The Garnet Family is Easy to Know

(although chemically it is the most complex of the minerals)

* The crystals are squarish with twelve or more small faces
* More often than not found in crystal form

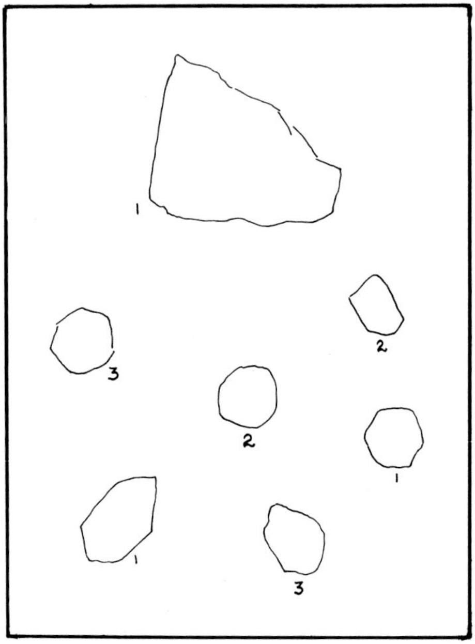

1. Cinnamon
2. Pyrope
3. Almandine

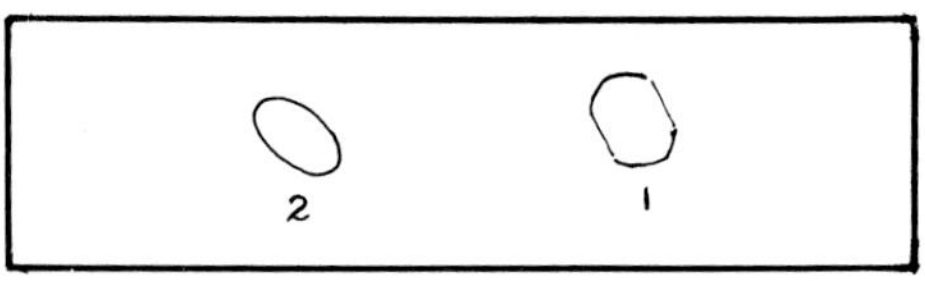

Gemstones
1. Pyrope (facetted)
2. Almandine (facetted)

The Feldspar Family is Easy to Know

(it is the commonest mineral of all)

* Breaks into flat faces which show a pearly luster as you move the specimen

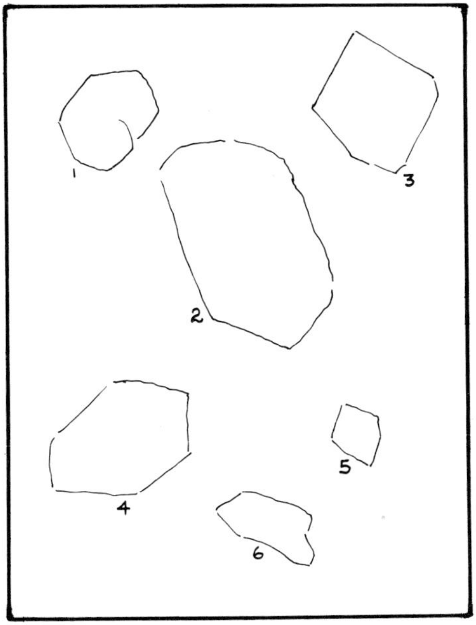

1. Twinned crystals of Orthoclase
2. Cleavelandite and Albite
3. Amazonite
4. Feldspar
5. Salmon
6. Labradorite

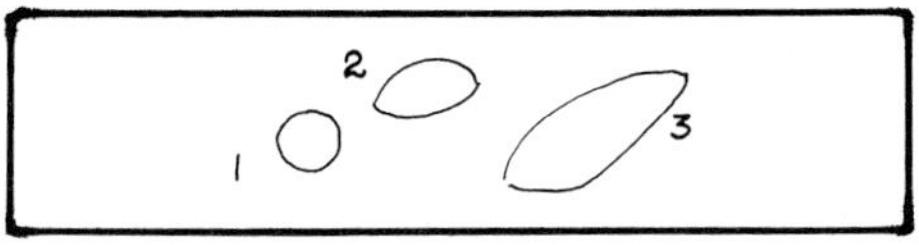

Gemstones
1. Moonstone (cabachon cut)
2. Pink Moonstone (cabachon cut)
3. Labradorite (cut freeform)

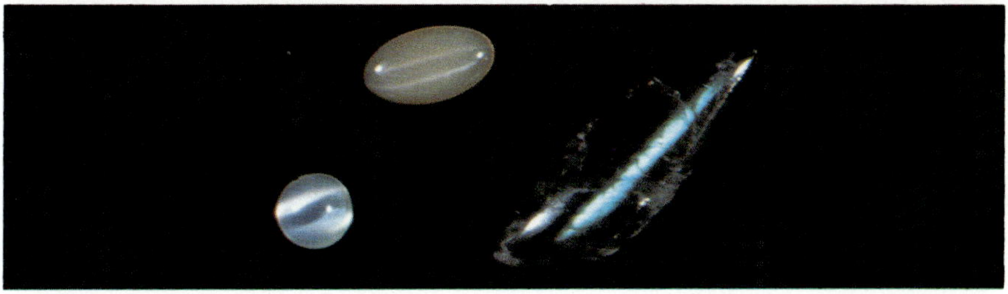

The Mica Family is Easy to Know

* No other mineral splits so perfectly into thin sheets

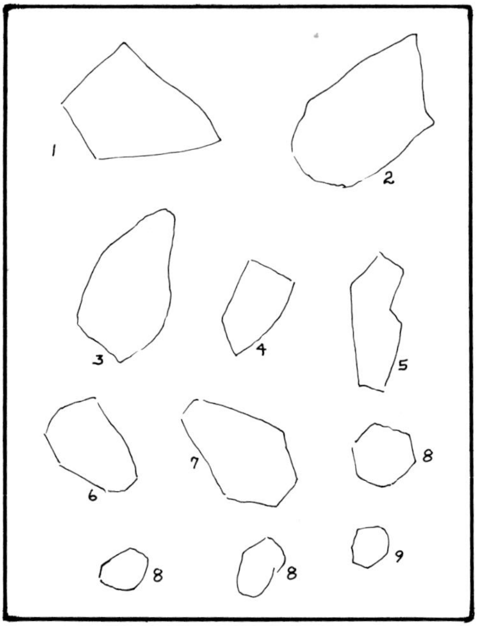

1. Biotite
2. Yellow
3. Ball Mica
4. Mica Schist
5. Curved Biotite
6. Lilac Lepidolite
7. Phlogopite
8. Muscovite
9. Muscovite crystal

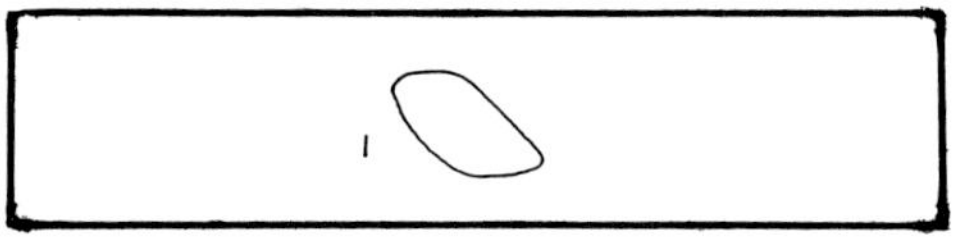

Gemstone

1. Lilac Depidolite (cut freeform)

The Metal Families are Easy to Know

* Dark colored
* Heavy
* The faces of the blocky little crystals shine like little mirrors on the surface of specimens

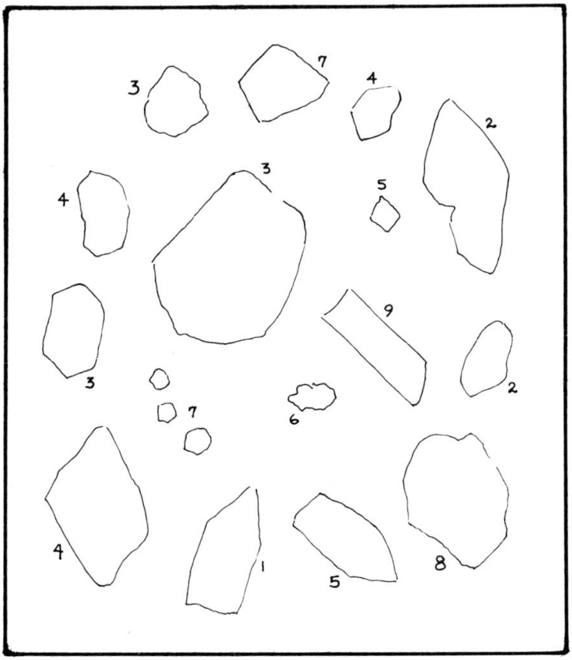

1. Pyrrhotite (magnetic iron)
2. Galena (lead)
3. Sphalerite (zinc)
4. Chalcopyrite (copper)
5. Silver
6. Gold
7. Pyrite crystals
8. Bornite
9. Core sample showing Chalcopyrite

"All the wonder and wealth of the mine in the heart of one gem"

Robert Browning

FURTHER IDENTIFICATION (Minerals)

The further identification of the six minerals (quartz, tourmaline, beryl, garnet, feldspar, and mica) we are considering is not as easy.

Quartz

Silicon dioxide SiO_2. Hardness 7. Gravity 2.6. Streak white. Found in every class of rock. Formed under all sorts of conditions. Next to feldspar it is the commonest mineral.

Uses: an important industrial material for glass manufacture and for making electronic equipment such as oscillators for controlling radio frequencies. Clear rock crystal, cut into spheres, is the gypsies' crystal gazing ball.

CHALCEDONY GROUP

Usually banded and translucent. Translucent red-orange is carnelian; apple green is chrysoprase; chestnut brown, yellow or reddish is sard; alternating black band is onyx; layers of light sard alternating with carnelian is sardonyx; gray, banding obvious is agate.

JASPER, CHERT, AND FLINT GROUP

Opaque and usually not banded. Dull brick red, yellowish, brown, grayish green, bluish is jasper; gray or brownish is flint; red is bloodstone or heliotrope. The name bloodstone comes from the old legend that this stone lay at the foot of the cross where drops of Christ's blood fell on it.

QUARTZ ENCLOSURES AS GUEST MINERALS

Tree-like forms in brown, red or black is mocha stone; long fine needles of red rutile in clear crystal is cupid's darts; same as above but in green is Thetis' hair; brown is Venus' hair; brown or green fern-like inclusions is moss agate; bar of light when cut cabachon is hawk's eye; bar of light slit lengthwise is cat's eye (chatoyant); bar of light in a tawny, lustrous yellow is tiger's eye; showing a star when cut cabachon is astriated.

Tourmaline

Maine's official gemstone. This playful mineral will try to confuse you. All in fun, however, for it really is a love. Chemically, it is the most complex of all. It has been called the harlequin of stones for centuries because it so often approximates gems from other families, such as rubies, sapphires, and emeralds. At times, exuberantly it shades from one color to another, then a third in the same crystal. Watermelon, and the newly discovered grape jelly from Newry are examples.

Luster glassy. Hardness 7-7½. Gravity 3.0-3.3 Fracture uneven to conchoidal. Cleavage poor. Transparent to opaque. Streak pale colored to blue-green. Strongly charged electrically on heating or cooling. Magnesium varieties (forming in dolomite) fluoresce yellow in short wave ultraviolet light.

The tourmaline series consists of:

SCHORL

$NaFe_3B_3Al_3(Al_3,Si_6O_{27})$ $(OH)_4$. Black. The most common and least valuable. Its crystals are largest. May be distinguished from black metals, such as columbite, because it is much lighter in weight.

SODA-DRAVITE

$NaMg_3B_3Al_3(Al_3Si_6O_{27})$ $(OH)_4$. Brown. Usually found in metamorphosed limestone.

LIME-DRAVITE

$CaMg_3B_3Al_3(Al_3Si_6O_{27})$ $(O,OH)_4$ White. The gem variety is achroite.

LITHIA-TOURMALINE

$Na(Al,Fe,Li,Mg)_3B_3Al_3(Al_3Si_7O_{27})$ $(O,OH,F)_4$. Tourmaline has two sprightly qualities. It shows dichroism, a color change when viewed in one direction, then another. Also, it becomes strongly charged electrically when heated or cooled.

The Dutch traders discovered this when a seafarer tossed a handful of the bright crystals he had brought home from afar to his children playing in a sunny courtyard. The delighted shouts of the little Dutch children brought the grownups to watch the sun warmed gems drawing after them bits of straw and courtyard litter. (Who knows, the seafarer may have picked up the tourmaline bits in the oldest mine in Maine, on Georgetown Island. Old-time sailors have told me it was regular practice to stop there before a voyage for a pocketful of the pretty baubles to carry around the world.)

Uses: Gemstones above all. The more glassy black varieties are used for electrical apparatus which depends upon their pressure—electric (piezoelectricity) characteristic.

Beryl

Beryllium aluminum silicate. $Be_3AI_2Si_6O_{18}$. Luster glassy. Hardness 8. Gravity 2.6-2.8. Fracture conchoidal. Cleavage poor basal. Streak white. Transparent to translucent.

One of the most beautifully crystallized of minerals, usually in prismatic hexagons. Its six-sided crystals sometimes occur very large, several feet in length and many pounds in weight. It is in Oxford County, Maine, where these giant beryls outdo themselves.

The high quality and large sizes of its gem varieties, especially aquamarine, have for years been putting Maine on the world's gemmological map. Beryl is usually found in pegmatites.

Uses: Gemstones. Ordinary beryl is the chief ore of the rare space metal, beryllium.

Garnet

Luster glassy. Hardness 6-7½. Gravity 3.5-4.3. Fracture conchoidal to uneven. Cleavage none. Transparent to translucent. Streak white to pale. Crystals squarish with four equal sides. Garnet, like tourmaline, usually occurs in crystal form.

This highly complex family is divided into six types.

Series 1 — aluminum silicates with magnesium, iron, and manganese.

Series 2 — calcium silicates with chromium, aluminum, and iron.

PYROPE

$Mg_3AI_2Si_3O_{12}$. Deep yellow-red. Found in igneous rocks, sometimes in volcanos.

ALMANDINE

$Fe_3AI_2Si_3O_{12}$. Deep violet-red. Found in metamorphic rocks, especially in mica schists.

SPESSARTITE

$Mn_3AI_2Si_3O_{12}$. Dark brown to black, reddish, and pinkish. Found in open cavities on metamorphic surfaces.

Series 2 — calcium silicates with chromium, aluminum, and iron.

UVAROVITE

$Ca_3Cr_2Si_3O_{12}$. Emerald green. Found in chromium deposits coating seams. The brilliant green gem garnet, demantoid, is perhaps the most rare and costly of all gemstones.

GROSSULARITE

Ca AI Si O. Various pale tints, not reds. Found embedded in metamorphosed limestones or seams or as crystals in pegmatite pockets. The light brown gem, essonite or cinnamonstone, is a grossularite garnet, as is the massive green gooseberry garnet.

ANDRADITE

$Ca_3Fe_2Si_3O_{12}$. Pale tints to brown and black, not reds. Found in igneous and metamorphic rocks on seams and in crusts, but not in mica schists.

Uses: Gemstones. Makes the abrasive sandpaper called garnet paper.

Feldspar

This is a group comprising orthoclase and microcline and the plagioclase series.

ORTHOCLASE

Potassium aluminum silicate. $KAISi_3O_8$. A potash feldspar. Hardness 6-6½. Gravity 2.6. Streak white. A common mineral distinguishable from another feldspar, spodumene, by its blocky, as opposed to splintery cleavage and from the plagioclase series by its lack of twin striations on the good cleavage face and by its 90° cleavages.

Colors: white, colorless; yellow and brown. A variety, sanidine, is light and glassy. When it reflects a bluish sheen, known as adularescence, it is the gemstone, moonstone.

MICROCLINE

Potassium aluminum silicate. $KAISI_3O_8$. A potash feldspar. Hardness 6. Gravity 2.5. Distinguished from orthoclase by its larger crystals and its occurrence in pegmatites, in which orthoclase does not occur.

Colors: white; flesh; red-brown and green. It is the only feldspar which produces the bright green variety, amazonite or amazonstone, used in jewelry.

THE PLAGIOCLASE SERIES

A continuous series of mixtures of sodium and calcium aluminum silicates. $NaAISi_3O_8$ to $CaAI_2Si_2O_8$. A soda feldspar. It is characterized by repeated twinning shown as many fine parallel lines on good crystal faces.

Albite

Gravity 2.63. Found in pegmatites in thin platey crystals. The variety, cleavelandite, was named after its discoverer, Professor Cleaveland of Bowdoin College in Brunswick, Maine. The costly moonstones from Ceylon and Madagascar are albite.

Oligoclase

Gravity 2.65. Bluish cast. The variety with reddish golden inclusions is the gemstone, sunstone.

Andesine

Gravity 2.71. Fine grained.

Labradorite

Gravity 2.71. Gray with a play of blue, green or bronze light across its surface like on a butterfly wing. It makes a beautiful gemstone and handsome and decorative building stones.

Bytownite

Gravity 2.74. Pale yellow. Found in lime rich igneous rocks.

Anorthite

Gravity 2.76. Rare. Found in lavas.

Uses: It has been used for centuries to glaze fine porcelain and in pottery and paint. It makes scouring powders and abrasives, too. When the demand slows, there is always chicken grit.

Mica

This mineral is possessed of the most perfect basal cleavage known. There are four varieties of mica:

MUSCOVITE

Hydrous potassium, aluminum silicate. $KAI_3Si_3O_{10}$ $(OH)_2$. Luster glassy to pearly. Hardness 2-2.5. Gravity 2.8-3. Streak white.

Colors: white, colorless, light yellow, amber, bright rose, and green.

The crystals look flat and hexagonal. They develop an interesting six-rayed star around the point of impact by a sharp point, known as a "percussion figure." They are found in all igneous and metamorphic rocks, in granite, schists, and in pegmatite dikes. Other minerals, such as tourmaline, are often trapped between the plates of the muscovite and grow in characteristic patterns of flattened crystals.

BIOTITE

Hydrous potassium, magnesium, iron and aluminum silicate. $K(Mg,Fe)_3AlSi_3O_{10}(OH)_2$ Luster glassy. Hardness 2.5-3. Gravity 2.8-3.4. Streak light to medium.

Colors: dark brown to black.

Biotite is a common mineral of pegmatites, especially those rich in rare earth minerals, for which it is a useful indication. The associated feldspar is often brick red.

PHLOGOPITE

Hydrous potassium, magnesium and aluminum silicate. $KMg_3AlSi_3O_{10}(OH)_2$. Luster pearly or metallic on cleavage face. Hardness 2.5-3. Gravity 2.7. Commonly asteriated.

Colors: light to dark brown.

Usually found in metamorphosed limestones and dolomites; sometimes in serpentines and igneous rocks. It can be told readily from muscovite by its association with crystalline marbles, its golden brown color (lighter than biotite) lesser transparency, innumerable microscopic inclusions. Because phlogopite often occurs in large sheets, it is in good demand for electrical purposes.

LEPIDOLITE

Hydrous fluosilicate of lithium, potassium and aluminum. $K_2Li_3Al_4Si_7O_2$. Luster pearly and vitreous. Transparent to opaque with a glittering fine grain. Hardness 2.5. Gravity 2.8-3.3.

Colors: lilac; gray-green and pale yellow.

Lepidolite is an important ore of lithium, the lightest of metals, found only in lithium-bearing pegmatites.

Uses: Once valued for windows in iron stoves and lanterns, isinglass, it is now much sought after as an insulator for electrical equipment. Scrap mica has many uses from lubricants to Christmas snow.

FURTHER IDENTIFICATION (Metals)

Besides gold, which we will honor with a separate section of its own, and silver, the four principal ore minerals found in Maine are in the chemical family known as sulphides. That is a combination of metal and sulphur. They are pyrite (iron), galena (lead), sphalerite (zinc) and chalcopyrite (copper).

These sulphides occur in or near the contact zone between where sedimentary (metamorphosed) have been intruded by igneous rocks.

The matrix in which the metals occur is very dark. It is likely to be ultra basic rock of the following dark, schistose intrusives: plagioclase feldspar, pyroxene, olivine and biotite, the black mica.

There is often a rusty surface of gosson (oxydized metal).

Pyroxene gives the matrix a bluish green color. This is the same "blue ground" in which diamonds are found in Africa, not Maine, though there are those who claim diamonds have been found in Maine, too. In Africa this blue ground is called "kimberlite."

Pyrite and Sometimes Pyrrhotite

PYRITE

Iron sulphide. FeS_2. Luster metallic. Hardness 6-6½. Gravity 5.0. Fracture conchoidal. Cleavage none. Often crystallized, most frequently in striated cubes, less commonly in pyritohedrons or octahedrons. Massive pyrite is common. Streak dark gray to black. Color light yellow. Pyrite is found in all classes of rocks and in all types of veins. It is associated with all types of metal ores. It is called "fool's gold" because it is so often mistaken for gold. It is an important source of sulphur and sometimes of gold.

PYRRHOTITE

Ferrous sulphide. Fe_1-S. Luster metallic. Hardness X. Gravity 4.6-4.7. Fracture subconchoidal. No cleavage. Crystals are usually tabular in thin hexagonal plates. Most commonly massive and granular. Brittle. Magnetic, a distinguishing characteristic. Color bronze.

Pyrrhotite is a common mineral of magmatic segregations and high temperature ore veins. It also occurs in pegmatites and in contact-metamorphic deposits. Sometimes contains the ore minerals, pentlandite (an iron-nickel sulphide) and chalcopyrite.

Meteorites contain the closely related, non magnetic, ferrous sulphide mineral, troilite.

Galena

Lead sulphide. PbS. Luster metallic. Hardness 2.5-2.7. Gravity 7.4-7.6. Cleavage perfect cubic. Brittle. Crystals very common, usually cubic, sometimes octahedrons with dull faces. It is found in medium and low temperature ore veins. Color lead-gray. Galena is the chief ore of lead. It is commonly associated with the sulphides, sphalerite, pyrite and chalcopyrite and with quartz, siderite, dolomite, fluorite, calcite or barite as worthless associates (gangue minerals). Occurs in both sedimentary and igenous rocks in veins.

Sphalerite

Zinc sulphide. ZnS. Luster adamantine to resinous. Hardness 3.5-4. Gravity 3.9-4.1. Fracture conchoidal. Cleavage perfect dodecahedral. Brittle. Transparent to opaque. Sometimes fluorescent, orange in ultra-violet light and shows the remarkable phenonenon of triboluminescence in that it emits flashes of orange light on being lightly stroked with a hard substance like a knife or a stone. Tetrahedral crystals are common. Also stalactitic, granular and massive. Color: colorless (rare) through yellow to red-brown and black. It is found in sulphide ore veins in all rock classes associated with pyrite and galena. It is the principal primary ore of zinc.

Chalcopyrite

Sulphide of copper and iron. $CuFeS_2$ Luster metallic, often with irridescent tarnish in brilliant hues. Hardness 3.5-4. Gravity 4.1-4.3. Fracture uneven. Cleavage one poor (and rarely noted). Brittle. The spenoidal crystals are common and often large. Also found massive. Found in sulphide veins and disseminated through igneous rocks. It is the most important copper ore. Color golden.

Silver

Silver. Ag. Luster metallic. Hardness 2.5-3. Gravity 10.0-11.0. Very malleable and ductile. Fresh surface bright white, usually blackened by tarnish. Forms in cubic or octahedral crystals, or more commonly in long distorted wires. Found in ore veins.

LESS COMMON METALS

Less common in Maine, but reported from three hundred fifty localities by the Maine Geology Department, are:

Nickel, scheelite (tungsten), manganese (source of iron), stibnite (antimony), bismuth, molybdenite ("white gold" — makes the finest steel for airplane construction), platinum, marcasite, arsenopyrite (arsenic), tellurium (tellurides of gold and silver), hematite, cassiterite (tin) and pentlandite (an iron-nickel sulphide).

THREE SPECTACULARS

Now — let's let ourselves get excited about these three spectacular, uranium, fluorescents and gold before we calm down in the next chapter and prepare ourselves for our treasure hunting adventures.

Uranium

This atom bomb metal occurs in nature only in combination with other elements as mineral compounds.

The host compounds are divided into two classes: primary, not changed since originally deposited, and secondary, formed by the weathering or other processes of the primary.

The primary uranium ore is uraninite. In the veins which cut through metamorphosed sediments and volcanics it is commonly of the botryoidal (kidney like) crust type known as pitchblende.

Secondary ores are very different in appearance. Instead of black they are characterized by bright coloration, yellow, orange and green.

Colors:

Bright yellow — carnotite
Orange and yellow — gummite
Lemon yellow — autunite
Green — torbenite

In Maine likely areas to wander over with a geiger counter are where pitchblende deposits are associated with sulphide ores and native silver, especially if hematite has left red stains in the vicinity.

Potash rich (perthite) pegmatites sometimes yield fine cabinet specimens and will drive the geiger counter, and you, crazy, but pegmatites are just too small and variable in their mineral content to justify a uranium mining program. Alas!

Uranium mineralization leaves signs of its radio-active attack for us to read.

Fluorite becomes dark, almost black and a "stinks-path," giving off an unpleasant smell when broken.

Celestite turns blue on bombardment.

Quartz becomes smoky.

Beryl turns golden.

Feldspars in the rare earth pegmatites are red-stained.

Allanite radiates cracks and sports a colored halo.

Hyalite opal and chalcedony become brilliantly fluorescent yellow-green when contaminated by uranium. Contaminated — ha!

Some claim that large fields of yellow flowers are such hopeful indications that it would be profitable to prospect from a low flying plane later to test drill where the patches of yellow were noted. I suggest that the pilot double check by dangling a scintillometer, the most sensitive and expensive detector of radioactivity.

Geiger counters are widely used in prospecting for radioactive deposits. To use a geiger counter, first find out the ground count of the area to be prospected. Then count the number of light flicks, buzzes in your ear and numbers recorded by the needle on the dial. Four and up beyond the field count are significant.

Uranium is a geologic time clock. Uranium atom after uranium atom transforms itself to lead, releasing helium. An analyses of the amount of lead of helium compared to the remaining uranium, in consideration of the known rate of radioactive decay, immediately gives the figure of the time which has elapsed since the mineral came into being in the place where it was found. The only weak point in these analyses, which give the earth an indicated age of at least five hundred million years, is an uncertainty about the possibility of a partial escape of some of the elements.

The locations which gave the strongest geiger counter responses in a U. S. Bureau of Mines survey of Maine were: Red Hill near Rumford, Oxford County; Melrose Quarry on the slope of Sugarloaf Mountain in Stoneham, Oxford County; on the slope of the Nubble in Greenwood, Oxford County; Noyes Mountain, Greenwood, Oxford County; the Ernest Wentworth prospect in Albany, Oxford County; Bumpus Mine in Albany, Oxford County; Black Mountain, near Rumford, Oxford County.

It's my guess that this survey was confined to the pegmatites of Oxford County because there is no mention of Hancock County and others where metal ore bodies abound and the vein formations are right for the presence of uranium.

Fluorescents

A sure fire parlor game, using a black light in a dark shadow box to make dull, ordinary looking rocks startle by vibrating with intense color.

Among the Maine minerals with the ability to glow with brilliant colors under an ultraviolet light are autunite, calcite, hyalite (opal) and opal coated feldspar, a variety of feldspars, manganapatite, litchfieldite, and from certain localities, zircon, tourmaline, and beryl.

Gold

Gold. Au. Luster metallic. Hardness 2.5-3. Gravity 19.3. Very malleable and ductile. Most often in octahedral crystals, but also in distorted clusters of parallel growths forming feathery leaves, wires or thin plates. Distinguished from metal sulphides by its softness and malleability. Streak gold. Color rich yellow to silvery yellow, depending on the amount of silver alloyed with it. It is found in quartz or sulphide veins and stream deposits. Its inertness and great density make it concentrate in stream beds either in small flakes or large nuggets, where it may be recovered by panning.

VEIN GOLD

Hancock County is where is found the greatest concentration of gold in veins in quartz or sulphides. In Chapter 5, I have listed seventy-three localities from this county and several others.

Elmer Hoyt, Falmouth mineralogist, told me of a mother lode of gold in the Small Point pegmatite in Sagadahoc County and of "black gold" — oil — in the sandstone and shale at the Portland Airfield in Cumberland County.

It is of wide occurrence giving rise to the witticism, "Gold is where you find it."

In matrix it occurs mostly in white quartz or in sulphide veins.

Once you have seen it either visible in vein quartz or gleaming through the black sand in the bottom of your gold pan, you won't mistake it. The late Martin Keith, owner of mines in Auburn and of a gold mine on the old Keith farm in Livermore, declared, "Once you get the color of gold established in your eye you will never go wrong on it."

Gold has been reported from over a hundred localities in Maine.

ALLUVIAL OR PLACER GOLD (found in sand or gravel in streams)

Gold was found first in Maine, not in 1854 in Perham Stream, Madrid, as has been thought, but before that in the crop of a chicken which had been pecking on the banks of the Swift River at Byron in Oxford County. The most likely places for gold are as follows:

Franklin County — Perham Stream, Madrid, Niles Brook, Rangeley, The Carrying Place, Carrabasset, Dallas Plantation, Sandy River Plantation, Haynes Township (5R6), New Sharon, Moscow, and Strong (Blue Ledge); Knox County — Appleton; Oxford County — The Swift River Byron, and Houghton; Somerset County — Skowhegan and Pittsfield; York County — Acton.

METHOD OF PANNING GOLD

Specifications: Gold Pan — 16" round, slope sided, tin pan, purchasable at one of the mineral stores in Maine. Sluice Box — 1' wide wooden trough narrowing towards the bottom, 6' to 10' long. The bottom may be lined with canvas. 18" long box at top has a very coarse wire-meshed bottom to strain out the larger rocks and rubble with a little gate to regulate, or stop, the flow of the water. Ladder-like riffle of 1" framework is placed in the bottom of the trough.

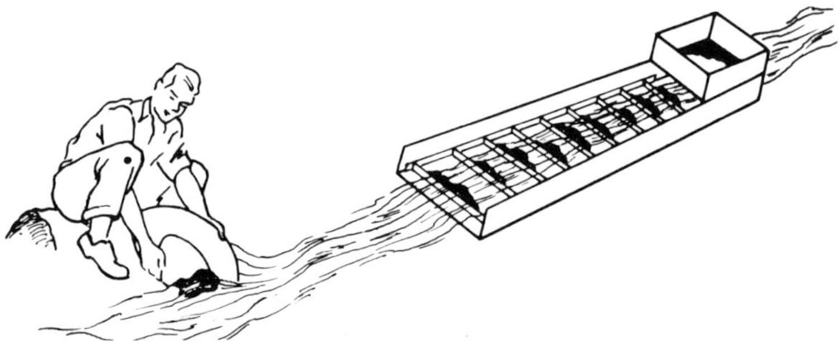

Where: Look for a natural catch basin in the river where gold is likely to be trapped. This would be where a bend in the river takes an inward curve and the bedrock riffles slant backwards towards the current.

Getting Material to be Washed in the Sluice or Panned: From the banks of the catch basin area, not from the water, where the spring freshets may have deposited gold, pry aside large boulders with a crowbar and dig with a shovel from the side towards the flow of the water.

Sluicing: Wedge the sluice box solidly between rocks in the river. Dump the rubble you have dug, into the small box off and on all day. Pour pail after pail of water into the box each time you have filled it and let the rush of the river water wash the residue down the trough. The heavier gold will settle behind the riffles and the lighter material wash away. At the end of the day, carry the sluice box ashore. Remove the riffle. Gently wash the residue into a gold pan and proceed to pan out carefully any gold which may have collected.

Panning: This is how to recover gold either from the residue in a sluice or from rubble freshly dug. Seat yourself comfortably on a rock where plenty of water rushes by. Grasp the heavy, rubble-filled, pan on each side and tip away from you. Dip gingerly in the river to let the flow slip off the lighter material. Lift and stir the material from the bottom upwards, then wash repeatedly for about ten minutes. The time will depend on your deftness as there is a flip of the wrist technique to it. When there is left in the bottom only red garnet crystals and sand, and black magnetite sand and, we hope, a flake or two of gold, exercise greater care. Again and again, allow a little water to swirl around, then to slip some of the sand out. When you get down to a small amount of black sand — STOP. Swirl a little water around the pan keeping a hawk eye out for gleaming yellow dots. "Now you see it — now you don't," so be alert to STOP. Force the frayed end of a wooden match down on the flake and transfer it to a small vial partially filled with water.

GOLD IN QUARTZ OR SULPHIDE VEINS

The Maine Geological Survey in the booklet it brought out in 1958, Maine Metal Mines and Prospects, reported gold in veins from these localities: Forty in Hancock County, One in Cumberland County, Three in Knox County, Two in Somerset County, One in Oxford County, Nine in Penobscot County, Five in Waldo County, Ten in Washington County, Two in York County.

Most of my experience seeking gold has been in the east branch of the Swift River where it rushes along the mountain above Byron in Oxford County. I was taught to pan and use a sluice box by the late Carl Skilling (Karl Schilling), a prospector who made his living for forty or more years from Swift River gold and to some extent from the five dollars an afternoon we tenderfoots gladly paid him for imparting his knowledge to us.

Others have made a living from Swift River gold.

The grandfather of a friend of mine set up a small gold recovering outfit in the river. His gold factory employed forty Italian workmen all with families to support. And he was enabled to raise and educate a fine big family.

Perley Whitney, a Lewiston jeweler who had panned in the Swift River for fifty years, claimed that more than twenty thousand dollars worth of gold had been panned from the river in his time there and the river is still yielding up its treasure, perhaps less prolifically than years ago, so many from all over are after it every summer.

Attempts to locate the mother lode are thought to have been fruitless. Yet Perley sometimes found gold in pockets in the watershed hills along the river. He took four hundred dollars worth from a single pocket.

At Houghton, just north of Byron the residents will point out to you the rusting crushers of the False Gold Mine.

In 1905 Timothy Boudreau and J. A. Delisle organized the Oxford County Gold Mining Company. They set up two heavy crushers, employed ten men and sold stock to the tune of a hundred fifty thousand dollars. The Federal officers put a stop to the venture with fines and jail sentences for the two swindlers.

Just beyond there at Rangeley another prospector used to make his living panning gold in Niles Brook. Jake the Trapper used to sell nuggets which were half gold and half platinum.

When anyone remarks to me, "Who knows. You might start a gold rush or a mining boom in Maine" I answer, "If I do it won't be the first one."

Gold rushes and wildcat gem mines, some legitimate, some fake and none profitable peppered the state from 1878 to 1883. While the fever was running high a monthly gem magazine was published in Maine.

One day I happened to be telling a lawyer in Augusta about the Gardiner ne'er do well who went the rounds of the businessmen showing the yellow stuff he had found in his newly acquired gravel pit. When a U. S. assay showed it to be gold the sale of stock in his gold mine represented the first folding money he had earned in years. The stock hit a bearish slump when they found he had salted the pit with flakes shot from his gun. "I guess I know that's a true story," the lawyer laughed, as he took from his wall safe his grandfather's certificate of the worthless stock. "Office safes all over Maine contain stock certificates as valueless as this."

It doesn't take much to touch one off. Enter another gold rush each time the witching yellow stuff shows up in a new spot.

Byron buzzed after the chicken yielded its golden crop.

In 1850 Henry Merryman was clamming on the larger of the Gosling Islets in Casco Bay. In turning over a chunk (junk) of clay he revealed a large gleaming yellow nugget. All the subsequent diggings of all the inhabitants of Harpswell Neck have so far failed to produce any more gold. Did an Ammoscogin Indian lose the nugget years before? Or had it been buried there by a PIRATE?

In 1854 the finding of gold in Madrid in Franklin County set the area afever.

Thirty years later Franklin County gold rushed once again when Cyrus Campbell, a white water lumberjack, found a several ounce nugget while hunting his lost pickpole head below Pondsayer Rapids in Lang Plantation.

Northern Franklin County just a few years ago yielded gold which assayed from $40.00 to $250.00 a ton.

About that same time young Herbert Thurston of Ellsworth in Hancock County picked up a shiny green rounded stone of abnormal weight. It kicked around the Thurston household for sixteen years until his father decided to throw it out. When he noticed that yellow particles of dust clung to his fingers he had it assayed. Young Herbert had brought home a seven pound gold nugget worth, at $35.00 an ounce, $3,920.00.

But such a minor event would never start a gold rush in Hancock County — not this most golden of Maine's counties!

Aroostook was heard from in 1895 when Richard Longley located at least five gold deposits with his homemade divining rod. Never let it be said that our Maine dowsers limit themselves to finding water! As a young man he had taken a fancy to a store boughten rod. But he didn't have its $150.00 price. So he made his own which for forty-three years showed him, and him alone, where either gold or silver lay.

Several years ago in Aroostook County at the head of Great Machias River west of Ashland a lumberjack stubbed his toe on a small but heavy rock. Cursing, he picked it up to heave it. Surprised at its weight he presumed it was gold. But — ah, no. Pure platinum.

Now — no more reminiscing around the cracker barrel!

We must get going.

We know how to recognize the twelve minerals and metals we are likely to meet. And we have learned to pan for placer gold.

A few pages of preparation and we'll be ready to take off.

4. Be Prepared

I have to tell you, though it hurts me as much as it does you, that you will find it MIGHTY SLIM PICKIN'S.

You will do most of your hunting in the mine dumps, excuse me — tailings — which are, after all, other people's leavings. And the miners have eyes, too.

Worse, there is no mining in Maine for gems. They are incidental to the getting out of commercial rock.

Worse yet, pocketty or otherwise gemmy areas are unwelcomed as impurities which interfere with the marketable value of feldspar and mica. Therefore, many a find of lovely gem material has been blown to smithereens by a heavy charge of dynamite.

One last echo from the voice of gloom. Hundreds of collectors have been there before you.

Had you better stay at home? Oh, no! Don't do that!

To quote once again John Allen May in The Christian Science Monitor when he wrote a second time about my having introduced him and his family to rock collecting, "It is a hobby without parallel . . . You may become rich but you are more likely to become happy."

Travel light when you do go. This is easier said than done!

WEAR what you usually would on a hike. Rugged slacks, warm socks and jacket and sturdy shoes. Canvas gloves and glasses with plastic safety lenses are a protection.

FOOD AND WATER as for a picnic only more of it.

FIRST AID KIT well to keep in the car. But, actually, you won't mind the jabs from quartz splinters and other minor casualties. You'll see.

THIS BOOK leave that in the car, too. Don't waste time in the field identifying. Bring home whatever looks pretty and interesting to look up at your leisure.

LUG these up the mountain:

Prospector's pick, a large and a small cold chisel, a two or three pound sledge hammer with a short handle (our Maine rocks are HARD!), shovel or clam hoe, short crow bar, compass, magnifying glass, flashlight, knife, a packet of paper bags, one for each specimen, pencil to write (on each bag name of mineral, date and from where collected), U. S. Department of Interior Geological Survey maps showing the quadrangles you are planning to visit.

Return light, too. Try to be selective with the specimens you bring back.

RULES OF THE ROAD

Ask permission. Somebody owns it, no matter how wild and remote. For each mine I shall state the owner but if you can't locate him inquire at the nearest house. If you still can't find anyone to give you the green light let your conscience be your guide.

Be tidy and in every way careful of the property.

If lucky, offer to buy your finds from the owner.

Keep well away from machinery.

Leave equipment alone. Bring your own tools.

Find out which are dumps you may explore and which are stockpiles of commercial rock.

Don't climb the walls of the dumps or the quarries.

Don't throw rocks over the edges of quarry walls.

WILL YOU BE WELCOME?

Yes, in most cases, if you indicate you are considerate.

However, you can scarcely help presenting a problem in a working mine.

When Dana Douglas was mining Bumpus, Albany in Oxford County for beryllium, a most serious and costly operation, he never dared blast without first counting rockhound noses.

Thought he, "If I post a sign, ROCKHOUNDS $5.00, that will end my problem."

Did it? No. He was embarrassed to death everytime someone walked up to him proferring a bill, "Do I pay my $5.00 to you?"

HANDY HUNTING HINTS

Head for the dumps. More chance if you go to lesser known locales where fewer rockhounds will have beaten you to it.

DIG! Expect to have to dig down two or three feet to reach pay dirt.

Dig around the roots of trees and bushes. Roots growing out of a ledge will sometimes lead right into a pocket. When Charlie Bragg shook the roots of an uprooted blueberry bush, topaz crystals fell off.

After a rainstorm the edges of a dump or the hill below it are likely areas.

Above all DIG!

CLUES IN PEGMATITE AREAS

A concentration of muscovite mica suggests the presence of beryl.

Lepidolite points to beryl.

An area is hopeful where things look all mixed up like a bread pudding. Especially is that so if perthite (an intimate intergrowth of quartz and feldspar) is present.

Where there is large blocky white feldspar there are usually no other minerals except large beryls, if in areas where they occur.

Spodumene is another hopeful sign on a tourmaline area.

Lavender mica is a tracer for tourmalines.

Cleavelandite is a clue to colored tourmalines. This dead looking, bluish white mineral with an interior shine sometimes outlines the hidden opening of a gem filled pocket.

Bits of colored tourmaline outcropping suggest more inside.

Gem tourmalines are generally found in pockets in pegmatites. Smoky quartz crystals often form the roof of the pocket with other crystals studding the walls or lying loose in the kaolin in the bottom. Most pocket tourmalines are green or pink of, roughly, three sided pencil form, sometimes well terminated. Before discarding the kaolin examine it carefully for gem crystals buried in it.

Aquamarines, colored tourmalines, and colored quartzes are to be found in pocketty areas of potash feldspar.

In matrix rich in soda feldspar are to be found topaz, fluorite, herderite, hamlinite, certain types of apatite and amethyst.

Thin veins of translucent to transparent quartz are apt to be host to aquamarine. The transparency of the quartz and the tendency of the potash feldspar and the mica to be in nice crystals sparsely distributed throughout the mass are good indications.

A vein of milky white quartz may contain pockets of clear quartz with occasional crystals of choice amethyst.

CLUES IN METAL ORE AREAS

Look for dark basic rock composed of plagioclase feldspar, pyroxene (pale bluish-green), olivine (dark yellowish-green), serpentine (green and greasy feeling), malachite (a wash of light to intense green associated with copper ore) and biotite (black mica). All these may not be present in any one specimen, but the general effect is dark.

Metal bearing rocks feel heavy.

The faces of small crystals scattered through the matrix will gleam like tiny mirrors.

Crystals, or veins streaking through the country rock, will shine with a metallic luster, varying in color depending on the metals present.

Color:

Gold-bright yellow. Sometimes in dark matrix, sometimes in white quartz

Silver — silver white

Copper, chalcopyrite-brass — yellow

Iron, pyrrhotite — bronze, sometimes oxydized to iridescent blues and greens.

Pyrite — brassy yellow

Manganese-iron — black

Lead, galena-lead — gray

Zinc, sphalerite-lead — gray

Iron rust, gosson (oxidized metal) will often show on the surface.

HOW TO GET OUT YOUR FINDS

Not so easy to get that choice crystal out of its hard rock bed!

Best you can do is to get it out in a large portion of its matrix and do the delicate job of removal at home, creeping up on it with a small chisel pounded with the hammer end of your pick. No guarantee of success!

Wield your sledge hammer well away from the specimen. Pry with the pick end of your pick or a crow bar. Protective glasses are recommended here.

HELP FOR YOU

There is no substitute for actually seeing and handling the minerals themselves and talking with people who know them well.

Here's where you can buy specimens, gemstones, jewelry, books, and equipment for prospecting, lapidary polishing, tumble polishing, and jewelry making.

Bragg, Charlie, Buckfield, Oxford County. An excellent guide.

Hasenfus, David, Boothbay Harbor, Lincoln County. He has just opened a new mineral shop.

Machias Jewelry Store. Robert Guptill, Machias, Washington County. He specializes in jasper from this area.

Marble, Charlie, Buckfield, Oxford County. A fine lapidary.

Maine Mineral Store, Trapp Corner, West Paris, Oxford County. Run by the widow and daughter of the late Stanley Perham.

Mann, Jim, East Bethel, Oxford County. Jim conducts his mineral pursuits from source to finish. He gets his material by hard rock mining. He cuts it into gemstones more original and artistic than any I have ever seen. He sells most to fine jewelers, but makes some jewelry himself. He will arrange to guide groups.

The Smiling Cow, Camden, Knox County. Owned by my sister, Margie Fisher.

Willis, Malcolm, Rock shop in Bar Harbor, Hancock County. One of the things he is noted for is his willingness to cut stones from your own finds. Many the fashionable dame who leaves her chauffeur with the limousine while she combs the shores for colorful stones for the Willises to cut. Her diamond tiara remains in the safe!

Winthrop Mineral Shop, Winthrop, Kennebec County. The owner is Priscilla Bryant. Her mineral background is bred in the bone. At four her playthings were bright stones from her grandfather Stearn's rose quartz mine on Paris Hill, Oxford County. At eleven she rigged up a laboratory in her Oxford County home where she analyzed chemically the endless variety of minerals Oxford County produced. Her Winthrop Mineral Shop has EVERYTHING you rockhounds could want. Every sort of specimen, jewelry, equipment to cut your own stones and to make your own jewelry, prospecting tools, books, and magazines. And here is the best place to buy tourmaline from the fabulous Newry find.

There are other rock shops throughout Maine which you can find by inquiring locally or by consulting *The Lapidary Journal Yearbook*.

Join Them And Go With Them

Members of Mineralogical Societies are generally knowledgeable and to go on field trips with them is a short cut to becoming expert yourself.

The Federation of Maine Mineral and Gem Clubs lists the following:

Kennebec Rocks and Mineral Club, Winthrop.
Lincoln County Gem and Mineral Society, Richmond.
Maine Mineralogical and Geological Society, Portland.
Oxford County Mineral and Gem Association, Rumford.
Penquis Rock and Mineral Club, Dover-Foxcroft.
St. Croix International Rock and Mineral Club, Calais.
Sandy River Mineral Club, Farmington.
Water-Oak Gem and Mineral Society, Waterville.

Museum Displays

There are good displays of Maine minerals at Colby College in Waterville, Bates College in Lewiston, and University of Maine at Orono and at Presque Isle. Bowdoin College in Brunswick has two of the most historic mineral collections in America, the James Bowdoin and the Cleaveland Collections. In the basement of Hubbard Hall is a cabinet of about 100 Maine minerals and a small fluorescent display. The Nylander Museum at Caribou has a fine fossil collection.

PROFESSIONAL MINING METHODS

Fifteen years ago the geologist, Don Wyke, created a stir when he prospected miles of pyrrhotite rich in nickel in Union, Knox County.

Although the present owners of the prospect are not now attempting to get out this vast store of nickel we can learn from the methods the previous owners, The Union Pyrrhotite, used in their complex and costly preliminary surveys.

Prospecting

Mining interests unfamiliar with the state's potentials can get information from the Maine Geology Department in the State House in Augusta. They had spent hundreds of thousands of dollars on preliminary surveys. They do this, as do private interests, too, in low flying airplanes dangling as many as three instruments. These are a magnetometer, a horizontal coil EM and a scintillometer. These record the presence of magnetic minerals, such as pyrrhotite, metals in general and radioactive minerals.

The state as well as private prospectors employ four methods on the ground.

1) Chemical soil sampling.

2) Use of a magnetometer, an instrument which measures the earth's magnetic field which is intensified by the presence of magnetic minerals.

3) Use of self-potential, an instrument which measures the electrical current in the ground which has been set up by oxidizing — sulphide ore bodies as a result of electro chemical action. This is similar to the action of a battery.

4) Geological methods.

Private consultants, too, are a source of information, such as Jack Rand, former State of Maine geologist who now has his office in Augusta.

Having pinpointed an area, the mining company obtains, for a small fee, permission to survey from each owner of the land.

If promising, mining leases are drawn up. Recently the state has passed very favorable laws as to mine leases on state owned land, copies of which they will send on request.

Mine Lease

One dollar cash in hand. A twenty-five year, sometimes shorter, lease guaranteeing to do an agreed upon amount of work a year. The land owners get 5% of the net. Sometimes they are paid an advance royalty which is later deducted from the 5%.

Union Pyrrhotite Survey

Then begins the preliminary survey. This will probably take five years to the tune of half a million dollars. About one hundred twenty thousand dollars a year are required to elevate the status of the prospect to the point where beginning to mine can be justified.

Step 1 — Area is gone over with instruments to determine the trend (direction and extent) of the ore body, which here is a sulphide of nickel, copper, iron and cobalt.

Step 2 — A survey grid is made thusly: A base line is surveyed parallel to and the length of the trend. Stations (stakes) are put down every fifty feet. Then, traverse lines are surveyed at right angles to the base line every two hundred feet with stations every fifty feet.

Step 3 — A professional reconnaisance uses eight different types of geophysical methods employing these instruments, at that time costing upwards of four thousand dollars each: AF MAG, Vertical Coil EM, Horizontal Coil EM, Self-Potential, Applied-Potential, Resistivity, Potential Drop Ration and Magnometer.

Details of Union Pyrrhotite

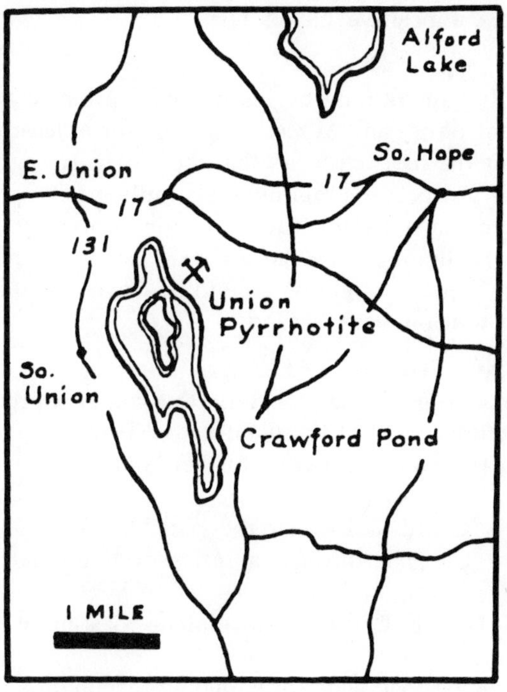

Step 4 — Diamond drills bring out core samples every two hundred feet, even when the ore body extends under water, as it does here under the state owned Crawford Pond and Alford Lake. The lowest grade samples were sent to the Union assay Office in Utah. They charged about twenty dollars for each five feet of rock from here, which is from two dollars to four for each metal present.

Roland F. Beers of Troy, New York, at that time head of Union Pyrrhotite, said that the drilling rigs set up on the ice of the pond the last two winters attracted a good deal of attention in the entire mining world. Also that the U. S. Bureau of Mines and the Colorado College of Mines made studies for this prospect resulting in discoveries of a new refining process for nickel. The pilot-type study was the first in the country. Finally, that this was the most serious long range mining exploration of the century in Maine.

Step 5 — Most of the metals present can be detected by the naked eye except nickel. This has to be determined by this test: dilute hydrochloric acid brings the nickel into solution. Ammonia converts it into basic upon which dimethylglyoxime powder can act. A resulting red color indicates a positive nickel test.

MINING

When at last the mining begins it should promise a minimum of six million tons of ore and expect to operate for at least twenty years. It is only after the first ten years that the company begins to make a profit. In this prospect the minimum six million tons are assured and a guesswork estimate is for fifteen to eighteen million tons.

Even in the mining, innovations were being planned. Gasoline driven mining vehicles might be able to be driven deep into the mine shafts and out again with a load.

EXTRACTING THE METAL

I can't hope to more than touch on this complex process, some of which will be done right at the plant before being shipped by coastal steamers and barges to refineries further south on the Atlantic seaboard.

Step 1 — Grind and crush the rock.

Step 2 — Carry it through a series of flotation and magnetic separations.

Step 3 — Next a fluidized-bed sulphate roasting of the bulk concentrate.

Step 4 — After that, water or ammonia carbon dioxide leaching to remove the nickel.

Step 5 — Finally, oxidation and precipitation of the metals.

AFTER THOUGHTS ON PROSPECTING

A relatively new and exciting system of electromagnetic prospecting is provided by one of those costly affairs, the AF MAG. This dispenses with a transmitter and utilizes a natural field source of energy. This field is primarily due to worldwide thunderstorm activity, often from over four thousand miles away. The electrical disturbance becomes concentrated in the ore body which acts like a radio antenna.

Most mineral shops and large hardware stores carry reasonably priced metal detectors, supposed to find metal as far down as fifty feet.

PEGMATITE MINING

Pegmatite mining in Maine is almost exclusively by quarry methods — open pit mining.

Mica mining at the Wardwell farm in Albany, Oxford County, will illustrate a different type of mining in Maine. It is during this type of operation that gemstork is sometimes unearthed.

A number of years ago a group from North Carolina, an area of pegmatites much like Maine's, were employing here the most modern methods of getting out high grade mica for use in electronics. These methods were also being used by Maine mining interests in the vicinity of Norway and South Paris.

Step 1 — A vein seven hundred and eighty-seven feet was surfaced out.

Step 2 — Three headings, forty-five to eighty-five feet long and up to thirty-five feet high were run off the main tunnel.

Step 3 — A crew of eight miners using the most modern equipment can average two tons a week.

Drills are of Swedish tungsten (we have tungsten here in Maine).

A dolly truck on a miniature railroad track, winched by steel cables, is powered by gasoline.

Step 4 — This hauled the mined mica to the surface and lowered supplies.

Step 5 — An automatic dump tripped its load into a waiting truck.

Underground, ten blast holes two feet apart are drilled vertically into the face of the heading. Each hole is six feet deep and large enough to handle a stick of dynamite one and a quarter inches in diameter.

A pneumatic hammer with chisel bit drill, the first used in the state, eats into the ground at a foot a minute.

Step 6 — The mica emerges from the blast in "books" (many-layered slabs), averaging six inches or more. One slab measured sixteen by twenty-four inches. Stan Perham got a six foot book of ruby mica out of his BB No. 7 mine! Another Maine "biggy."

Step 7 — It is then bagged, hauled to the top, unbagged, cleaned of rocks and is ready to be shipped to the depot.

Step 8 — It streamed out of the Albany mine at the rate of twenty-five hundred pounds daily to a shop in Norway where it was cobbed (sorted to eliminate the scrap mica which is shipped to Penacook, New Hampshire, for wet grinding), rifted (sheeted), then finally trimmed to the thickness of 0.007 to ⅛ of an inch. Mica trimming is a skill which in both Maine and North Carolina has been handed down from father to son.

Step 9 — The trimmed pieces, from one to three square inches, were packed and shipped weekly to Franklin, New Hampshire, to go into the U. S. government stockpile.

The federal government used to share the cost of mining strategic metals and minerals and participate in getting it out.

I am mightily impressed with the magnitude of hard rock mining, aren't you?

So — let's shoulder a shovel, grab a pick and GET GOING!

5. Off To The Mines

Beware of being misled by my "Oh, me. Oh, my. Stars in the eyes" treatment of the mineral wealth of the counties described in the following pages.

My facts are very little embroidered but there seem to be "ifs, ands, and buts" to many of them which do alter the finished fabric.

Truly a noble thing — the New England conscience!

For years I have wondered why Maine isn't jumping with profitable mining enterprises, in view of the bright prospects I keep hearing about. I asked our State of Maine geologist, Bob Doyle, about that. For instance, the manganese found in Aroostook.

"A dead duck," said Bob, whom I can always count on as an anchor to windward. "One hundred per cent! India has three times as much with plenty of cheap labor to get it out."

So, with a grain of salt for each, we proceed to the mines all over the State of Maine.

At the beginning of each county, the map designates the U. S. quadrangles and the townships where the mines occur. The mines are located by numbers and mine symbols, crossed geologists' picks. Highway route numbers, the smallest of dirt roads and mileages will direct you.

The text following describes each mine numbered on the map.

ORGANIZED ROCK HOUNDING

There just isn't any!

I tried to convince Bob Doyle to have the Geology Department present a plan to the mine owners such as:

1) The State provide an outfit: compressor, bulldozer, dynamite or whatever for each of the most likely collecting areas over the State. This outfit would go the rounds of the mines in the vicinity putting in fresh blasts turning over the dumps to expose new material. Engage guides.

2) Set fixed fees for the mine owners and guides. Establish necessary rules and insurance regulations.

3) Maine Development publicize this plan nationally.

The owners of the Murfreesburg, Arkansas diamond pipe have a workable plan found to be mutually satisfactory. They charge a fee for each person over twelve years old (under twelve have to wait to grow up). The entrance fee permits rockhounds to keep any ordinary finds. Each agrees to split fifty-fifty with the owner the value of anything over five carats.

I stressed the benefits to rockhounds, mine owners and the State.

When I finished my impassioned pitch, Bob grinned indulgently. "The idea would develop a three million dollar a year industry in the State," he said. "We have worked out a plan this long" (arms spread a yard apart). "We haven't come up against a stone wall. More like a plump, soft pillow."

Although the Geology Department has not made any further attemps, I KNOW it can be done. Because, I did it! In four days I led a group to eight mines in four counties. I arranged dynamiting to be done while we were there, hiding behind trees, at Blanche Bennett's two mines on Paris Hill in Oxford County. In Hebron we climbed up in the dark of night to the Upper Rubellite mine with my portable ultraviolet light to see the six different colors of fluorescent minerals light up the mine.

DIRECTIONS?

If my maps of the counties do not succeed in taking you to the mine of your choice, do not fondly suppose that, as you near it, someone is sure to be able to direct you. For some unexplainable reason, the nearer you get the fewer have ever heard of the place.

Should you stumble on one who does know, he may decide not to tell you, either to be perverse or to avoid helping you locate the valuable minerals he means to get for himself someday when he gets around to it.

It happened to Helen Caldwell Cushman and me as we clawed our way up Old Speck Mountain to Screw Auger Falls and beyond to Mother Walker Falls.

The Blue Paint Trail

The half mile a warden had told us would take us to the path to the falls had stretched to a mile, then two.

Before us a trail liberally spattered with sapphire-blue paint led up the mountain.

"This must be it!" So, up we went until it ended at a crazy little cabin with no sign of habitation. So, down we came again.

An amused fire warden told us it was just as well the owner happened not to have been at home. The woods queer hermit is convinced he has valuable minerals on his land — not so — and the blue paint trail is a spider web leading AWAY from his treasure.

Even if directions are supplied you with the best of intentions it is no guarantee you can follow them. How will you recognize the abandoned farmhouse landmark when a rich New Yorker has bought it and remodelled it? The tumble-down barn where the dirt path starts up the mountain has completed its tumble and been burned up in the parlor stove.

Don't ask the lone one deep in the woods, bent double in an intent search. He is an herbalist looking for the long list of medicinal herbs which bring fancy prices from the drug houses.

Here is a hopeful exit note, though, before we go on to the mines listed next, county by county, and located on each county's map.

The large paper companies have opened up some of their vast holding to rockhounds. The guards at the gates will tell you if you are welcome and perhaps give you directions.

You are welcome on State owned land.

ANDROSCOGGIN COUNTY

Here is part of the rich pegmatite sheet which in Oxford County yields a fantastic variety of minerals. Many notable gems have come out of the Androscoggin mines: aquamarines, pink cesium beryls, and tourmalines of the first water. Fancy gem smoky quartz is found here.

The mines are in typical soda-feldspar ledge.

Lewiston Quadrangle

LISBON TOWNSHIP

Map 1 . . . *Lisbon Paint Mine.* Still in existence from years ago are stock certificates for the "Lisbon Paint Mine." The paint is a strange dirt of various brilliant colors. It was first discovered in old Indian graves. They used to use it in many ways and found it would last forever.

ANDROSCOGGIN COUNTY

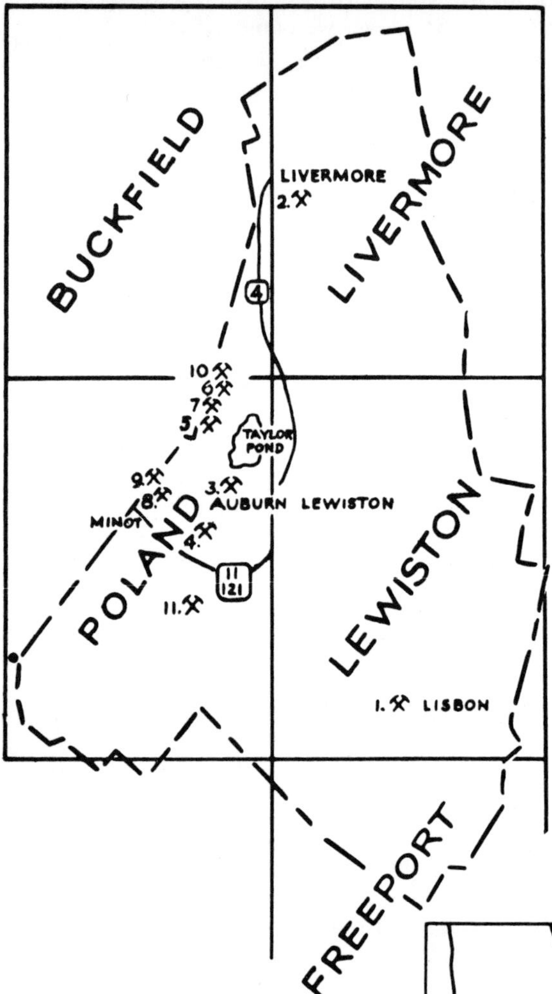

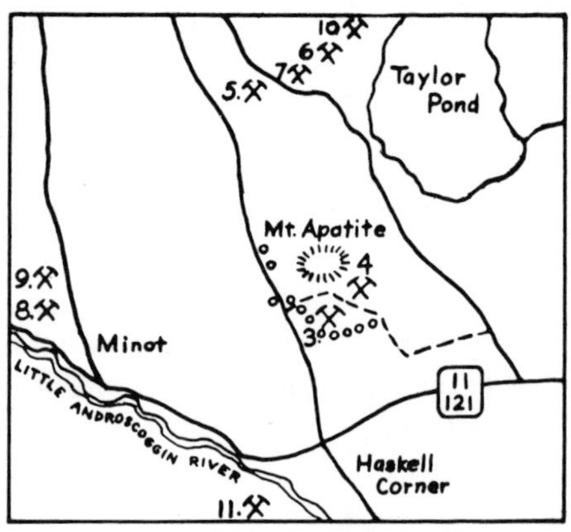

THE MOUNT APATITE MINES

Livermore Quadrangle

Livermore Township

Map 2 . . . *Keith Farm.* Martin L. Keith, gemcutter of Auburn, used to tell about a gold mine he helped work in the old Keith farm on the road from Bretten's Mill to Strickland Ferry in Livermore. A vein of gold went down fifty feet.

Poland Quadrangle

Auburn Township

Map 3 . . . *The Mount Apatite Mines.* owned by John Powers, Orono.

Androscoggin Tourmaline Co., Greenlaw (see below), *Hatch Farm, Keith* (Towne), *Maine Feldspar Co., Pit Pulsifer.*

The late Ulysses Greenlaw, former owner of the Greenlaw Mine, was as famous for his accomplished profanity as for his find of a twenty-four foot long crystal of deep pink cesium beryl (morganite).

It was Ule's habit to carry around in his pockets a fortune in magnificent gems cut by his friend, Martin Keith, in his day the finest lapidary in the State.

Freeman Burr, the then State geologist, thinking to benefit both Ule and his associates at the Capitol persuaded him to show them in Augusta.

After he had sold about $75.00 worth from his pockets, he became bored and returned home.

The next day the air was blue with his curses. " . . . I'll never, never go there again! Why the . . . should I sell any of my gems? I like them myself, else why the . . . should I carry them around in my pocket? Anyway, it's all too . . . much trouble!"

Map 4 . . . *Kennedy Mine.* Owned by B. F. Kennedy, Jr., 99 York Avenue, South Portland. Reported: beryl (aquamarine), black tourmaline, mica, garnet and quartz.

MINOT TOWNSHIP

Map 5 . . . *Bell Prospect (Giddings Prospect).* Reported: feldspar, muscovite mica and black tourmaline.

Map 6 . . . *LaFlamme Mine.* Reported: pollucite, feldspar, black tourmaline, beryl, purple apatite, garnet, lepidolite and spodumene. At one time, but not now, this mine used to support an industry. A geiger counter goes wild here.

Map 7 . . . *Phillips Mine.* Reported: autunite, feldspar, quartz (amethyst), and black tourmaline.

Map 8 . . . *Pitts Mine.* Formerly owned by Fred Pitts, Minot. Reported: feldspar, muscovite mica and quartz.

Map 9 . . . *Pitts-Tenney Mine.* Formerly owned by Fred Pitts, Minot. Reported: beryl, diopside, feldspar, and grossularite garnet. This cinnamon garnet is exactly the color of its name.

Map 10 . . . *Sturtevant Mine.*

POLAND TOWNSHIP

Map 11 . . . *Berry Quarry.* Owned by W. Levin, R.F.D. 4, Auburn. Reported: apatite, beryl, cassiterite, lepidolite, feldspar (microcline), muscovite mica and tourmaline. I have some beautiful deep green tourmaline crystals from here.

AROOSTOOK COUNTY

This whole potato empire in northern Maine is dotted with the iron bearing metal, manganese.

Over a million dollars have been spent by commercial, State and federal agencies to study mining possibilities here. No profitable procedure for handling the low grade (only 10%) Maine material has been devised for current market use.

Such high hopes! Now a dead duck, alas! Well, it's all here when we do need it.

Ashland Quadrangle

ASHLAND TOWNSHIP

Map 1 . . . *Frenchville.* Reported: manganese.

Caribou Quadrangle

PERHAM TOWNSHIP

Map 2 . . . *Perham Church, Bugbee-Bragdon, Fox-McIntire and Spaulding School.* Prospected by The Manganese Co., M. A. Hanna Co. Reported: manganese.

NEW SWEDEN TOWNSHIP

Map 3 . . . *Gelot Hill.* Reported: manganese.

Map 4 . . . *Haines, Aroostook River and Silver Mine.* Reported: silver and a large body of manganese.

AROOSTOOK COUNTY

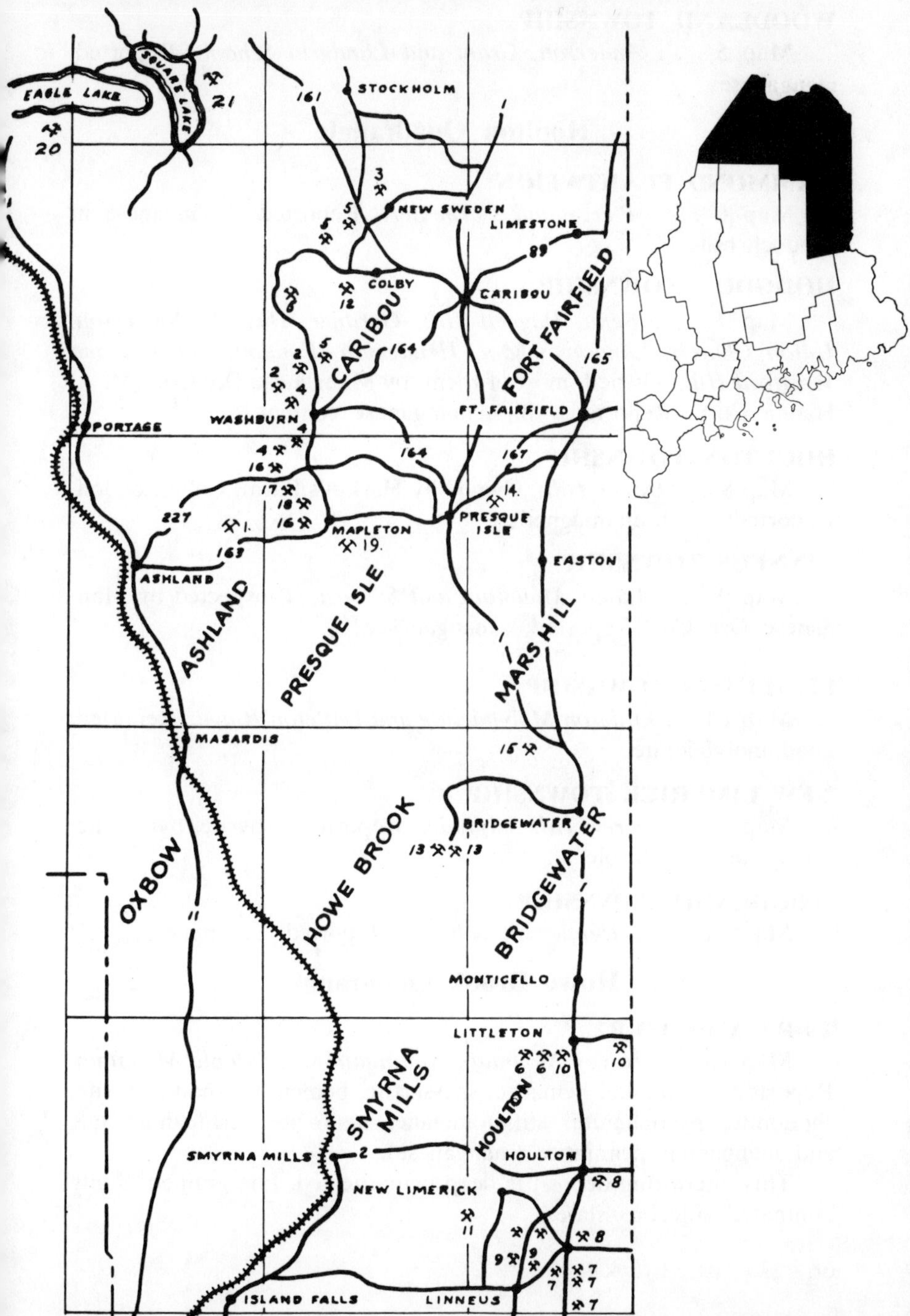

WOODLAND TOWNSHIP

Map 5 . . . *Anderson, Grant and Lundgrin School.* Reported: manganese.

Houlton Quadrangle

HAMMOND PLANTATION

Map 6 . . . *Pierce and Carpenter.* Reported: manganese in rhodochrosite.

HODGDON TOWNSHIP

Map 7 . . . *Benn, Daggett Hill, Gardner, Haskell, Nickerson, Union Church, London School, Henderson, London, Tourney and Westford Hill.* Owned, most of them, by Manganese Ore Co., (M. A. Hanna Co.). Reported: high manganese content.

HOULTON TOWNSHIP

Map 8 . . . *Hovey Hill.* Owned by Harkin's Farm, Colie London. Reported: rich in magnetite.

LINNEUS TOWNSHIP

Map 9 . . . *Adams Mountain and Stewart.* Prospected by Manganese Ore Co. Reported: manganese.

LITTLETON TOWNSHIP

Map 10 . . . *Littleton Molybdenite and Littleton Ridge.* Reported: good molybdenite.

NEW LIMERICK TOWNSHIP

Map 11 . . . *Drew Hill Sulphide.* Reported: pyrite, pyrrhotite, marcasite and marble.

WOODLAND TOWNSHIP

Map 12 . . . *Henderson School.* Reported: manganese.

Howe Brook Quadrangle

T9-R3 AND TD-R2

Map 13 . . . *Hovey Mountain continuous with Maple Mountain.* Reported: braunite, hematite, spessartite, bementite, rhodochrosite, rhodonite, pyrophanite, stilpnomelane, neotocite, manganian talc and manganian pennintite (manganese).

This interesting deposit is deep hematite red, fine grained, thinly laminated siliceous slate.

In 1951, the U. S. Bureau of Mines bulldozed 30,600 feet here and completed 25,809 feet of diamond drilling in 61 holes. They estimate a reserve of 256,000,000 long tons of 8.87% manganese ore of which 121,000,000 tons average 11.16% manganese.

Mars Hill Quadrangle

PRESQUE ISLE TOWNSHIP

Map 14 . . . *Maysville.*

WESTFIELD TOWNSHIP

Map 15 . . . *Reese.* Reported in 1957: manganese.

Presque Isle Quadrangle

CASTLE HILL TOWNSHIP

Map 16 . . . *Castle Hill Grange Hall.* Metal reported: manganese.

Dudley. Metals reported: braunite, bementite, manganiferous carbonate, and collophane (manganese-iron).

First prospected by the State of Maine, then by the owners, Manganese Ore Co., it was estimated to total more than 21 million tons, running 11.4% manganese and 20% iron.

Map 17 . . . *Richardson.* Metal reported: manganese.

Map 18 . . . *Smith.* Metal reported: manganese.

On the whole, these Castle Hill mines are manganese rich and are controlled by The Manganese Ore Co.

MAPLETON TOWNSHIP

Map 19 . . . *Higgins.* Metal reported: manganese assaying over 10%. Prospected by The Manganese Ore Co.

This is the location where many years ago the former State of Maine Geologist, Freeman Burr, first found this iron-bearing metal, manganese. It is useful, too, for hardening steel.

Some of the material at the Higgins Farm and the nearby Forest M. Dudley Farm assays 32.8%.

Eagle Lake Quadrangle

DEBOUILLIE TOWNSHIP

Map 20 . . . *On Debouillie Mountain.* Owned by the Great Northern Paper Co. Reported: molybdenum and copper. Not allowed on property.

Map 21 . . . *Square Lake, south shore*. Owned by The International Paper Co. Reported: copper and zinc. Not allowed on property.

Paul Zinner, former Chief of the Minerals Division of the U. S. Bureau of Mines, had established the Aroostook County material at 250 million tons containing 94% manganese.

He had succeeded in getting a government appropriation of $342,727.00 for pilot plant testing.

He stated that this Maine ore deposit, treated by the modern Nossen process, would go far to relieve the steel industry of dependency on foreign sources for 90% of its manganese needs.

CUMBERLAND COUNTY

Westbrook, Windham Center, North Windham and Standish in Gorham, Gray and Buxton quadrangles respectively, comprise the southern end of an unusual mineral sheet.

In micaceous schist is found staurolite, the "fairy cross" and andalucite, some of it pink, in quartz veins and some of the variety chiastolite, which we call "daisy chiastolite" because its cross crystallization forms a flower-like pattern.

This formation begins at a ledge by a bridge over the stream in Windham Center and extends for miles all the way to the Rangeley district and Kingfield in Franklin County.

The mineral sheet has yielded, too, colored tourmalines, spodumene and kyanite. The pretty, blue-bladed kyanite is used in paint manufacture and gave its name to the brand, "Kyanize."

In Cumberland County, on the Gosling Isles in Casco Bay, was where, over a hundred years ago, Henry Merriman, while clamming, dug up a large nugget of gold.

In Raymond, on Sebago Lake, in the Gray quadrangle, is Pulpit Rock. About fifty feet up from the highway is where the Devil used to stand when preaching to the Indians. When a rash young chief had the temerity to laugh in His face the Evil One rose in a passion and, stamping his feet in fury, caused the ground in front of the pulpit to drop fifty feet.

Nathaniel Hawthorne used to spend many happy hours reading on this rock.

Near the entrance to The Sebago Lake State Park is a large gravel pit. Standing in the center of the floor of this pit is a large boulder of

CUMBERLAND COUNTY

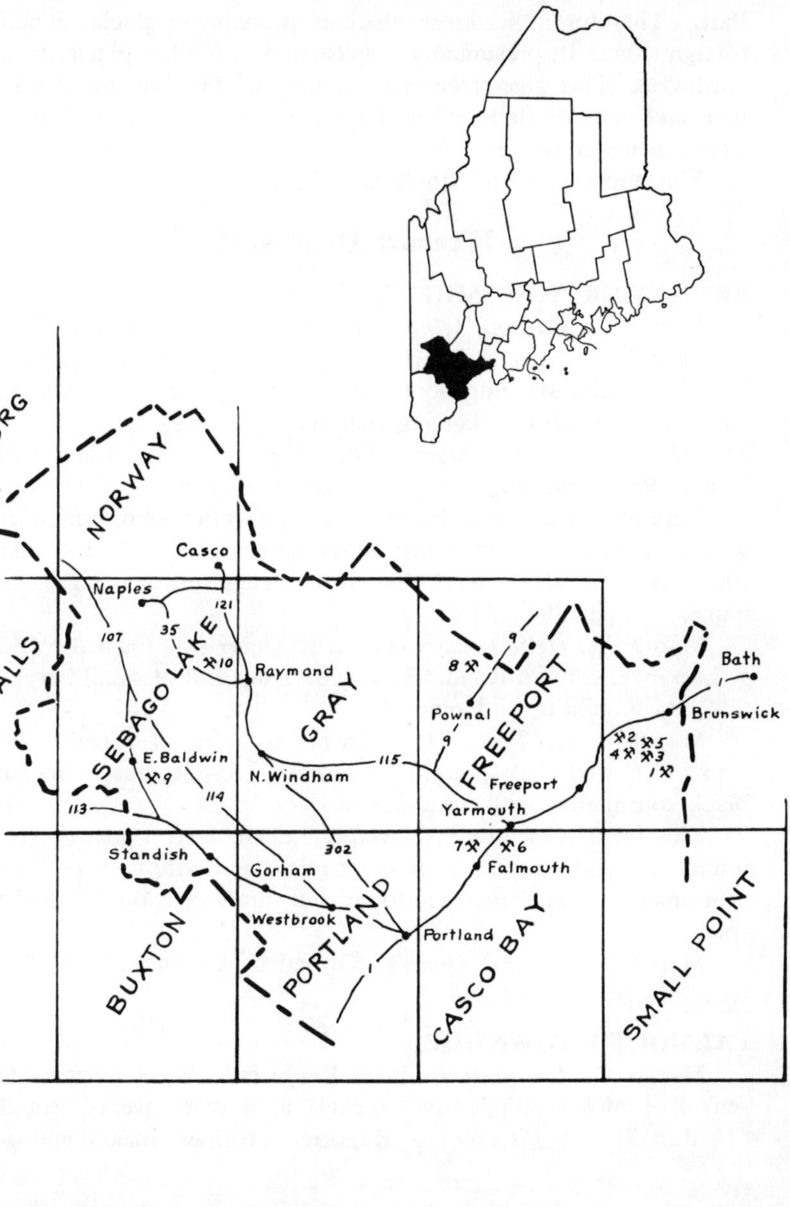

dark gray rock boldly streaked with veins of white quartz. It measures 5x10x12 feet and weighs at least 27 tons.

This metamorphic argillite rock stands out like a sore thumb because it is the only mass of argillite in the bedrock of the entire State Park. Therefore, this large block is probably a glacial erratic, or foreign stone. It, presumably, was carried by the last glacier from the northwest. This glacier carved out here a 200 foot circular kettle-hole and cobbles flattened and scratched by grinding against other stones beneath the glacier ice.

Fluorescent hyalite (opal) is found at Pownal.

Freeport Quadrangle

BRUNSWICK TOWNSHIP

Map 1 . . . *Garland Mine.* Owned by Walter C. Garland, Woodside Road. Reported: beryl, feldspar, muscovite and quartz.

The granite sheeting here is very perfect and nearly horizontal, so is in demand for flagging and underpinning.

Map 2 . . . *Grant Quarry.* Owned by Henry L. Grant, Hillside Road. Reported: feldspar, muscovite and quartz.

Here have been found hornblende crystals (a host of uranium) and graphic granite, that intriguing intermixture of quartz and feldspar which looks for all the world like ancient cuneiform writing, hence its name, "graphic."

Map 3 . . . *Helie Quarry (Larrabee Quarries).* Owned by Albert J. Larrabee, Church Road. Reported: ledge full of small beryl crystals. Also, biotite and quartz.

Map 4 . . . *LaChance Mine.* Owned by Jeffrey LaChance, Brunswick. Reported: beryl, biotite, columbite, feldspar, muscovite, quartz, black tourmaline and white topaz.

This favored locality produced one of the largest gem beryls ever found. It yields the finest of sheet mica, herderite crystals, translucent emerald beryl, fine blue topaz and fluorescent opal-coated feldspar.

Map 5 . . . *Morse Quarry.* Owned by Charles Morse, Church Road. Reported: feldspar.

FALMOUTH TOWNSHIP

Map 6 . . . *Falmouth Copper.* Reported: copper pyrites and arsenate of nickel (copper and nickel) in a shaft twenty feet deep.

Map 7 . . . *Piscataqua.* Reported: copper, nickel and gold.

POWNAL TOWNSHIP

Map 8 . . . *Tryon Mountain Quarries.* Owned by Philip Tryon, Pownal or Consolidated Feldspar, Topsham. Reported: biotite, feldspar and muscovite.

Sebago Lake Quadrangle

BALDWIN TOWNSHIP

Map 9 . . . *Kelley Prospect.* Owned by John Kelley, Jr., Hiram. Reported: biotite, muscovite, feldspar and quartz.

CASCO TOWNSHIP

Map 10 . . . *Chute Prospect.* Owned by Chute Bros. Lumber Co., South Casco. Reported: feldspar.

FRANKLIN COUNTY

There is considerable mineralization in this county. In the northern portion, which borders Canada, there are valuable deposits of copper, scheelite (tungsten) and gold, some of which assays from forty to five hundred and twenty dollars from a ton of host rock.

In 1886, Cyrus Campbell, lumberjack, found a several ounce nugget of gold below Pondsayer Rapids in Lang Plantation, Kibby Quadrangle.

An old legend centers around the Rich farm in Jay, Farmington Quadrangle. An early settler, John Rich, befriended an Indian who kept him plentifully supplied with huge chunks of pure lead to melt down for bullets. He would never reveal where he got it. To this day you can see the plain granite headstone marking the grave where Squire Rich buried his friend.

And it is handsome granite, pure white. The large blocks in Grant's Tomb, on Riverside Drive, New York, are from the quarries on the hillside in North Jay.

Farmington Quadrangle

FARMINGTON TOWNSHIP

Map 1 . . . *Farmington and Farmington Falls.* Metal reported: scheelite (tungsten).

KINGFIELD AND JERUSALEM TOWNSHIPS

Map 2 . . . On the Carrabassett River banks, red jasper is plentiful. This is the chief Maine locale for brecciated jasper.

FRANKLIN COUNTY

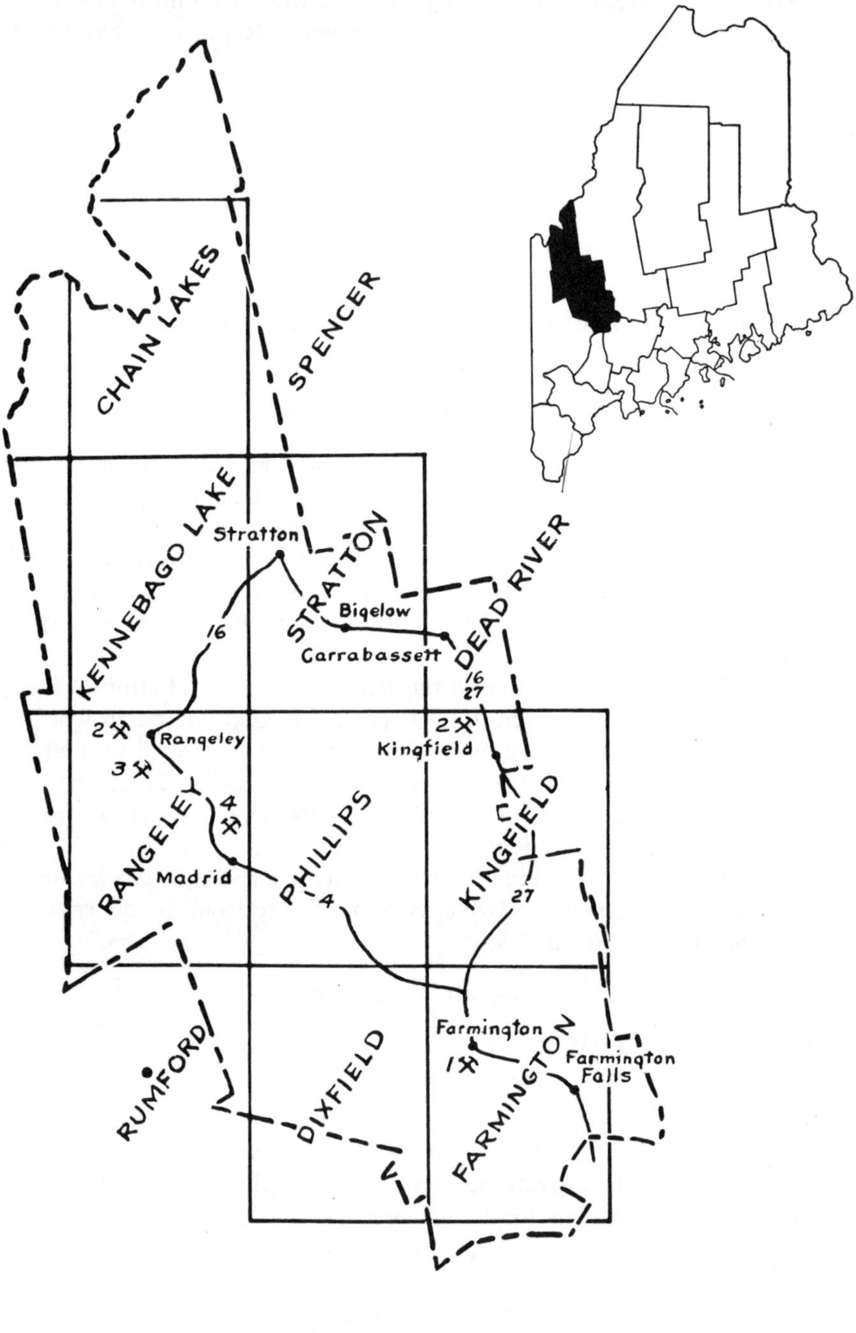

Rangeley Quadrangle

This is the northern extremity of the sheet of mica schist containing andalucite and staurolite, the southern end of which begins in Cumberland County.

RANGELEY TOWNSHIP

Map 2 . . . *Wing.* Reported: pyrite, pyrrhotite and calcite.

TOWNSHIP D

Map 3 . . . *Mountain Brook.* Reported: scheelite.

TOWNSHIP E

Map 4 . . . *Township E.* Owned by Maine Mining Bureau or Sidney Harden, Rangeley. Reported: columbite, feldspar (microcline), muscovite, quartz, spodumene and tourmaline.

HANCOCK COUNTY

Our streets are paved with gold!

Literally true, say the construction workers who see the glinting yellow flakes in the gravel they haul from here for the highways.

Hancock is THE metal county, including gold in veins reported by the State Geology Department from forty different locales.

In 1957, a large manganese deposit was found here by the State and the U. S. search planes flying at five hundred feet dangling a magnetometer ("doodle bug," our State Geologist calls it).

The manganese is widely distributed throughout the county as are antimony, pyrite, marcasite and pyrrhotite. Some platinum, iridium, gold and silver are found in most of the lead and zinc localities.

At Blue Hill, Beddington, Phillips and Houlton are found in bogs and ponds, diatomaceous earth.

Stone for the Panama Canal came from Mosquito Mountain about six miles up the Penobscot River from its junction with Penobscot Bay.

Acadia National Park Quadrangle

BAR HARBOR DIVISION

GOULDSBORO TOWNSHIP

Map 1 . . . *Clerque and Pol, Globe.* Reported: galena (lead) and silver in a shaft forty feet deep.

HANCOCK COUNTY

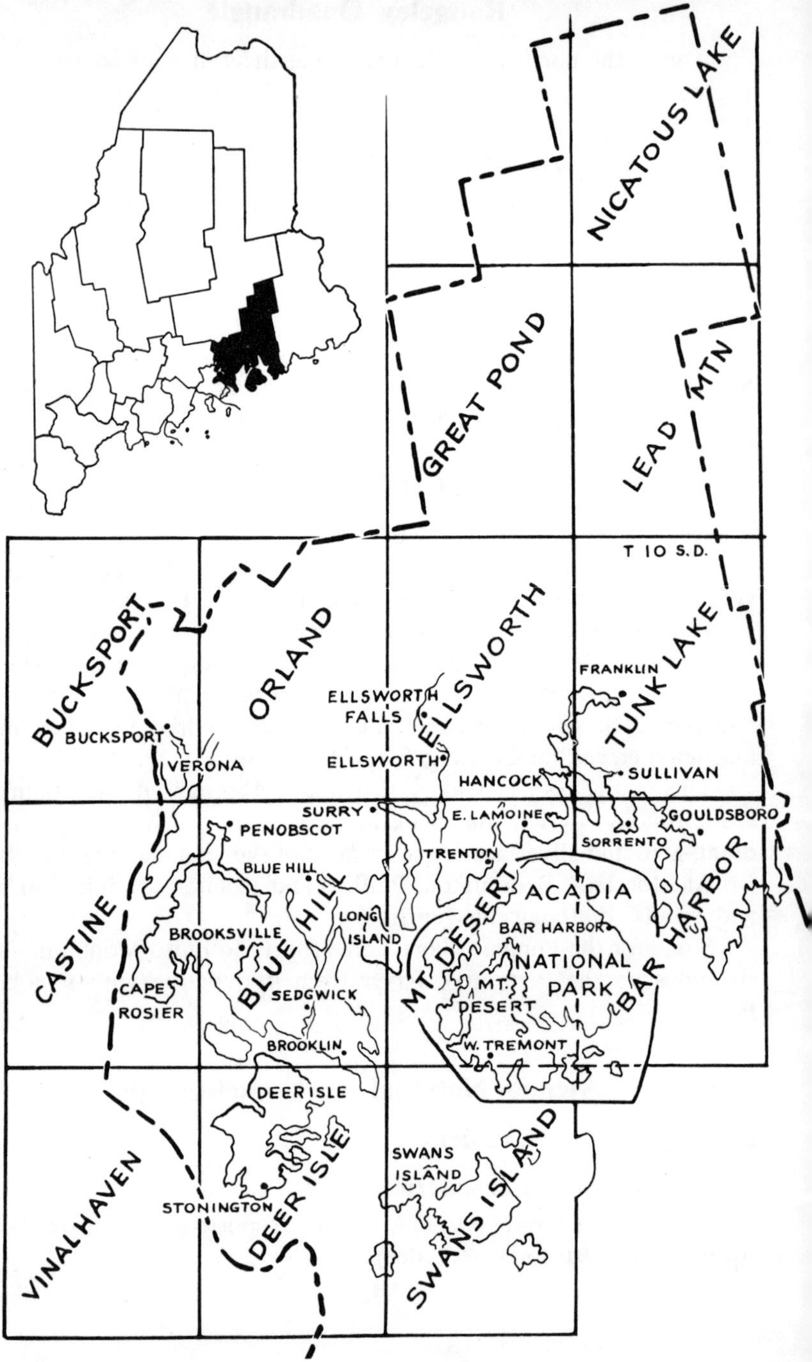

Gouldsboro. Reported: galena, sulphurets (copper, lead, silver, antimony, gold-trace) in a shaft 143 feet deep. Pay streak 2½ feet wide.

Hector. Two shafts on the Gouldsboro vein.

Homer. Reported: lead and copper. Two shafts. Considerable copper and galena at 80 feet depth in No. 1.

Kerana. Reported: sulphurets and carbonates of copper, galena, zincblende and iron pyrites.

Victor. Reported: lead and silver.

West Bay. Reported: located on the Gouldsboro vein. Shaft 60 feet deep is timbered to 17 feet; 8 feet of overburden. Vein is 13 inches wide at 24-foot level. Eighty ounces per ton of silver is reported for this vein.

West and Soule (Soule and West). Reported: shaft 70 feet deep on the Gouldsboro vein.

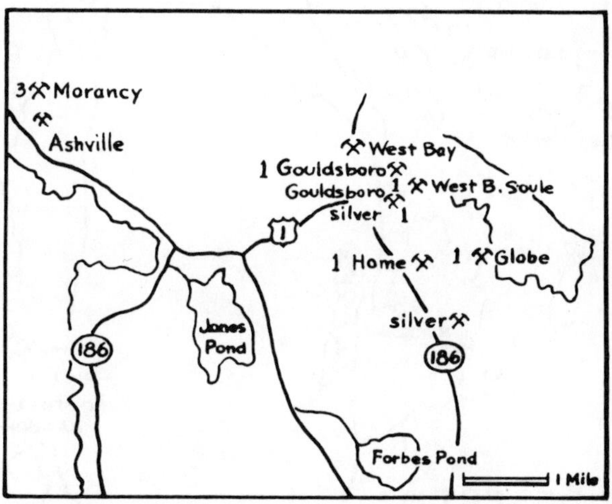

BAR HARBOR QUADRANGLE-GOULDSBORO AREA

HANCOCK TOWNSHIP

Map 2 . . . *Grant.* Reported: copper, lead and silver. Shaft 100 feet deep. Quartz with disseminated ore minerals highly mineralized with silver.

McFarland. Reported: argentiferous galena, chalcopyrite (lead, silver, copper traces and indications of gold).

SORRENTO TOWNSHIP

Map 2 . . . *Boss o' the Bay.* Reported: native copper in a twenty-five foot vein.

Golden Circle or Silver Isle. Reported: tellurides, silvanite (gold, silver, bismuth and tellurium). Gold quartz vein. Two compartment shaft 35 feet deep.

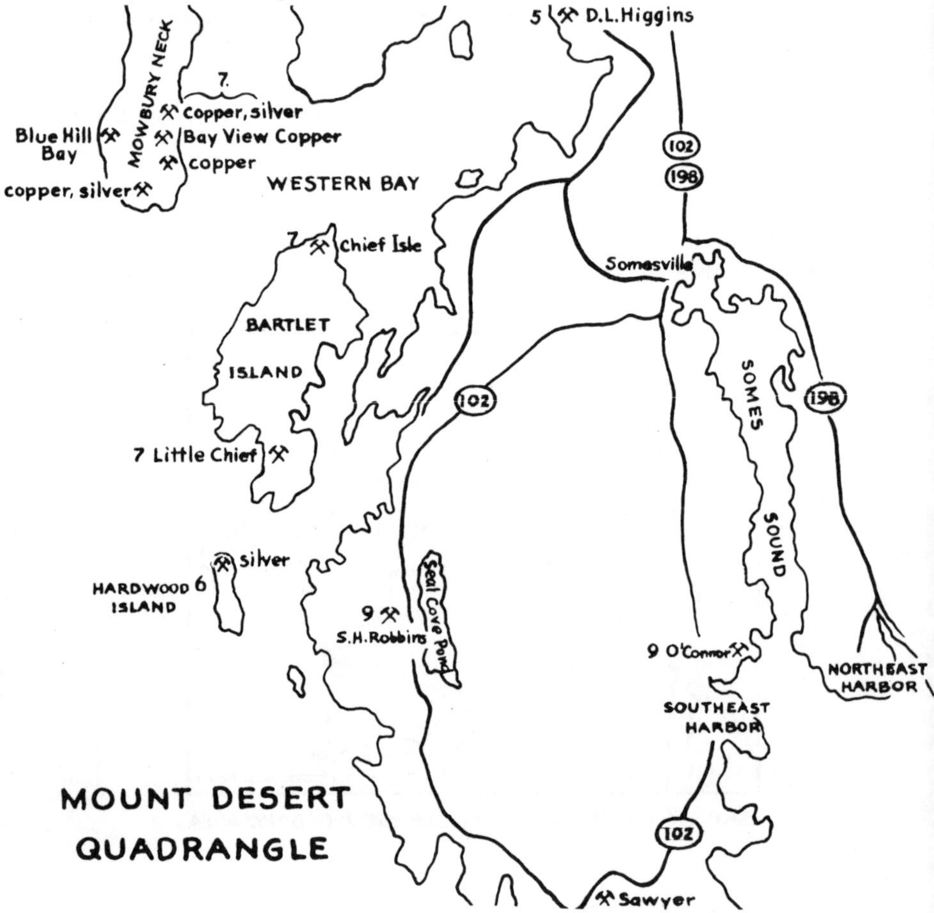

SULLIVAN TOWNSHIP

There is pure silver at Sullivan. In fact, this area has almost every metal and many minerals.

Map 3 . . . *Copper Harbor.* Reported: copper and gold.

Morancy. Reported: copper and silver in shaft 40 feet deep.

MT. DESERT DIVISION

BAR HARBOR TOWNSHIP

Map 5 . . . *D. L. Higgins.* Reported: copper. Out on the breakwater are amazonite (green feldspar), some in crystals, associated with red feldspar.

LAMOINE TOWNSHIP

Map 2 . . . *Eureka.* Reported: copper and silver in a shaft 40 feet deep.

Ford, Little Sue. Reported: five shallow prospect shafts show mineralization for a distance of 800 feet. Fine galena found in two shafts; vein 5 feet wide at 25 foot level in one of two.

MT. DESERT TOWNSHIP

Map 7 . . . *Chief Isle.* Reported: silver.

Little Chief. Reported: silver.

SURRY TOWNSHIP

Map 4 . . . *Blue Hill Bay.* Reported: copper and silver.

Curtis. Reported: copper and silver in a vein extending across the peninsula from shore to shore.

East Surry. Reported: vein of copper, silver and gold.

Happy Thought, Oakland. Reported: silver.

TREMONT TOWNSHIP

Map 6 . . . *Clark Gold.* Reported: gold.

O'Connor. Reported: tin.

S. H. Robbins. Reported: copper.

Romer. Reported: copper.

Tremont. Reported: silver, gold and copper.

Blue Hill Quadrangle

What a favored metal area! It abounds in arsenic, copper, gold, lead, nickel, silver, tin and zinc, with accent on the copper.

There are fifteen or more shafts at Blue Hill from which several million tons of copper have been obtained.

The Douglas Mine in Blue Hill township, No. 8 on the map, now Canadian-American owned and called Kerramerican, has an impressive list of metals with copper leading the parade. Beginning in 1883, when it was first opened, until 1957 when Texas Gulf Sulfer diamond drilled it, this two compartment, 350-foot shaft has been opened and shut regularly.

Before 1900, about three million pounds of copper were taken out. George W. Weed, minerals expert on the Maine Planning Board, said that in 1918 the Douglas mine produced five hundred thousand pounds of copper, enough to supply the State of Maine with license plates for five years.

The main shaft is 350 feet deep with ten tunnels, the longest of which is a mile in length and runs most of the way under a small pond, making this portion State land. The colossal Kerramerican operation is now taking out a thousand tons a day of copper and zinc and processing it at their own plant. They regret they cannot welcome visitors.

This is only a partial list of the sulphide and sulpho-salt groups of metals which have been found in the Blue Hill District: anglesite, argenite, arsenopyrite, azurite, bornite, brochanite, cerrusite, chalcocite, chalcopyrite, chrysocolla, cuprite, cuprotungstite, dioptase, enargite, galena, magnetite, marcasite, millerite, native copper, niccolite, pyrite, pyrrhotite, pyrargyrite, specularite, sphalerite, stibnite, tennantite, tetrahedrite, wolframite.

BLUE HILL TOWNSHIP

Map 7 . . . *Arctic (or Little Blue Hill).* Reported: copper.

Bay View (and Bay View Copper). Copper from 30-foot deep shaft.

Blue Hill Manganese. Reported: rhodonite (manganese).

East Blue Hill. Reported: galena (lead, silver and gold) from two shafts 34 feet deep.

Eclipse. Iron, copper, silver and lead from a vein 25 feet wide.

Map 8 . . . *Atlantic.* Reported: pyrites (copper, silver and gold). Two shafts, one 72 feet deep at which depth it cuts the Douglas lode.

Bisbee. Copper and silver in a shaft 55 feet deep.

Blue Hill, Blue Hill Bay and Blue Hill Central. Three shafts, deepest 180 feet. Pyrite and chalcopyrite (copper, silver and gold). Ruins of concentrating plant, smelter and slag dump on the surface.

City of Boston. Copper.

Darling. Copper, lead and silver.

Douglas — Now *Kerramerican.* Reported: pyrite, chalcopyrite, pyrrhotite, galena, arsenopyrite, sphalerite, magnetite, cuprite, stibnite, tetrahedrite, tennantite, molybdenite and native copper (cop-

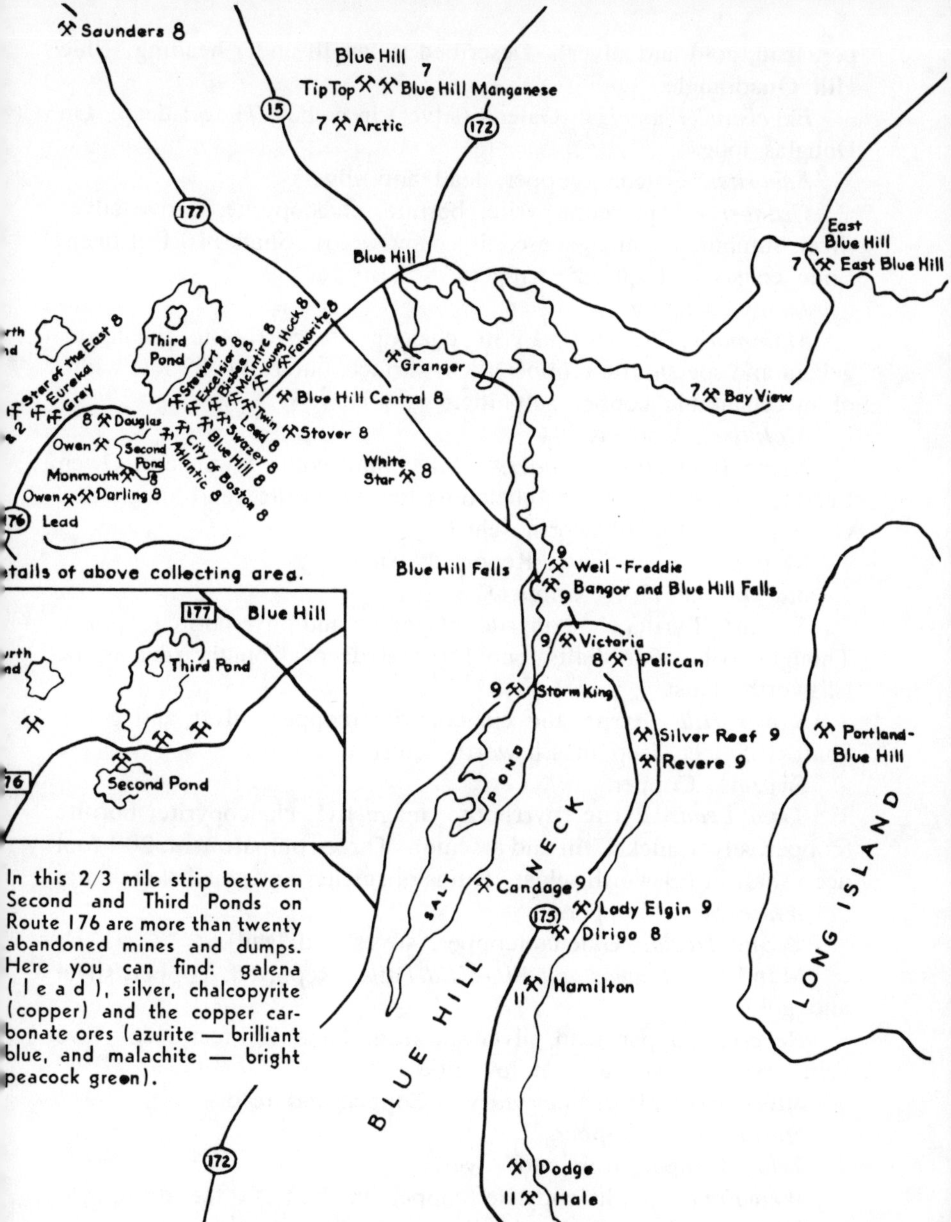

In this 2/3 mile strip between Second and Third Ponds on Route 176 are more than twenty abandoned mines and dumps. Here you can find: galena (lead), silver, chalcopyrite (copper) and the copper carbonate ores (azurite — brilliant blue, and malachite — bright peacock green).

BLUE HILL QUADRANGLE

per, lead, gold and silver). Described at length under heading "Blue Hill Quadrangle."

Excelsior (Phoenix). Galena (silver) in a shaft 71 feet deep. On Douglas lode.

Favorite. Galena (copper, lead and silver).

Granger. Reported: pyrite, bornite, chalcopyrite, native silver (iron, sulphur, tin manganese, silver and gold). Shaft 110 feet deep. Lode consists of quartz stringers.

Jersey. Copper.

Mammoth (Blue Hill). Pyrite, chalcopyrite, pyrrhotite, sphalerite, galena and specularite (copper). On surface, outcrop is 37 feet width of merchantable copper sulphide.

McIntire. Copper.

North Blue Hill (Saunders). Silver and gold. Owen and Owen Lead. Reported: pyrite, chalcopyrite, pyrrhotite, and magnetite. Open pits in the Ellsworth schist.

Map 8 . . . *Pelican.* Reported: silver.

Star of the East. Copper.

Stewart. Pyrite, chalcopyrite, magnetite and phyrrhotite (copper). Thought to be four shafts, one 150 feet deep. Country rock is the Ellsworth schist.

Stover Hill. Pyrite and chalcopyrite (copper, silver and gold). Shaft 105 feet deep in Ellsworth schist.

Swazey. Copper.

Twin Lead. Pyrite, pyrrhotite, magnetite, chalcopyrite, bornite (copper, silver, nickel, tin and arsenic). Three compartment, 200-foot deep shaft in Ellsworth schist. Ruins of smelter and small slag dump.

White Star. Copper.

Young Hecla. Galena (copper, silver and lead).

Map 9 . . . *Bangor and Blue Hill Falls.* Reported: copper, silver and gold.

Revere. Copper, lead, silver and iron. Shaft 130 feet deep. Two veins exposed on shore at low tide.

Silver Reef (Welcome Mary). Copper and lead.

Storm King. Copper.

Trio. Copper, iron and silver.

Weil-Freddie. Chalcopyrite (copper) in shaft 100 feet deep.

Victoria. Silver and gold.

Map 10 . . . *Candage.* Copper.

Lady Elgin. Copper, lead and silver.

CASTINE and BLUE HILL QUADRANGLES

Map 10A . . . *Portland-Blue Hill.* Reported: very good copper, molybdenum, and beryl.

BROOKLIN TOWNSHIP

Map 11 . . . *Brooklin.* Reported: gold and silver.

Carter. Copper.

Hale. Copper.

Hamilton. Copper.

BROOKSVILLE TOWNSHIP

Map 12 . . . *Blodgett.* Reported: copper.

Manhattan. Copper sulphurets, quartz (copper, silver and gold).

Shepardson. Copper.

Tapley. Pyrite, pyrrhotite and chalcopyrite (copper). In 1948, the U. S. Bureau of Mines diamond drilled five holes for a total of 1,204 feet. Two other shafts and an open cut indicated on Bureau of Mines map.

Map 13 . . . *G. Y. Mills.* Reported: copper.

J. P. Tapley. Copper.

T. C. Tapley. Copper.

Map 14 . . . *J. S. Douglas.* Silver.

Map 15 . . . *Chester.* Copper sulphurets.

Kenny Grindle. Owned by J. W. Davis and Sons. Reported: sulphurets of copper and iron.

PENOBSCOT TOWNSHIP

Map 16 . . . *Bagaduce.* Reported: copper.

Eastern Bell. Gold and silver in quartz.

Emma Albina. Copper

Map 17 . . . *Hercules (Dunbar).* Reported: galena, sphalerite, chalcopyrite, pyrite, pyrrhotite, specularite, native copper, native silver, arsenopyrite ("gray silver"), and copper silicate (silver, copper, zinc, lead and gold). Vein from 8 to 12 feet wide, exposed only at low tide. 80-foot shaft. Cofferdam was erected for initial development.

Menton Hill. Silver.

Mermaid. Silver.

Penobscot. Quartz crystals and carbonates (silver) in shaft 50 feet deep.

Silver Lake. Silver.

SEDGEWICK TOWNSHIP

Map 18 . . . *Eagle Brook*. Reported: lead in vein running parallel to Eggemoggin vein.

Eggemoggin. Pyrite, arsenopyrite, chalcopyrite, galena, sphalerite, native silver (lead, silver, zinc, copper and gold trace). Two shafts, one 120 feet deep were struck by lightning and burned in the early history of the mine. Ruins of an old metallurgical plant stand nearby. Host rock is devitrified rhyolite breccia of the Castine formation.

There is a gold mine and a silver mine good for collecting at Herrick's Landing. You go down the Castine Road to Herrick's Landing. Turn left into Sedgewick at the fork with the sign reading "Brooksville." The mines are located about three miles from South Brooksville, near Sargentville, just off the road to Deer Isle Bridge. The ore body contains some nice crystals of galena (lead) and sphalerite (zinc).

SURRY TOWNSHIP

Map 19 . . . *Sunburst*. Reported: galena, bornite, chalcopyrite, pyrite, malachite, and red oxide of copper (copper, silver, gold and lead).

Bucksport Quadrangle

BUCKSPORT TOWNSHIP

Map 20 . . . *Bucksport*. Reported: lead.

Comstock. Lead.

Eastern (Great Eastern Silver). Lead.

Rich. Gold.

Snow. Lead, nickel, tin, silver and gold. 45-foot vein is in quartz and quartzite.

PROSPECT TOWNSHIP

Fort Knox. Copper, lead, silver and gold in a 70-foot shaft.

VERONA TOWNSHIP

Empire. Lead.

Castine Quadrangle

Here is the quadrangle where The Fundy Bay Copper Co. of Toronto, Canada, acquired two thousand acres and in 1958 established The Penobscot Mining Co. to explore for copper, lead, zinc, manganese and whatever. This is around the Cape Rosier area at Goose Falls, across the Bagaduce River from Castine. It used to be owned by The Fundy Bay Copper Co., Charles Robbins, Jr., Newark, N. J.; John McComb and A. F. Thomas, N. Y., and F. C. Knight, Toronto, Canada, then by The Callahan Mining Company. We used to visit them time and again, don hard miners' hats and be browsed around their vast operation and through their processing plant by a knowledgeable geologist. No more. It is shut down because the ore body in the mouth of a large upper silurian volcano finally gave out, after they had extracted over two million tons of copper, zinc and lead. The pit was over three hundred eighty feet deep. An enterprising young man is now raising coho salmon in the large lagoon formed when they dammed the tide water.

From West Brooksville to Castine is a large body of zinc.

Reported mineralization: basic country rock, diorite; altered talc; chlorite; calcite; pyrite (iron); sphalerite (zinc-reddish or black-streaks red); chalcopyrite (copper); silver; galena (lead; pyrrhotite (magnetic iron ore. A small amount throws off their machinery).

BROOKSVILLE TOWNSHIP

Map 21 . . . *W. E. Tapley*. Reported: copper.

Map 22 . . . *Bakeman, Brooksville*. Copper.

Commonwealth. Copper.

Dodge (Jones and Dodge). Argentiferous galena, zincblende, copper sulphurets (copper, silver).

Mansfield. Copper. On Cape Rosier.

Rosier Extension. Zinc and copper.

G. E. Smith and G. F. Smith. Copper.

CASTINE TOWNSHIP

Map 22 . . . *North Castine*. Reported: galena, chalcopyrite, pyrite (lead, copper, silver and gold). 100-foot shaft with outcrop on the shore.

Map 24 . . . *Castine (Castine Head)*. Copper, lead, silver and gold.

Emerson. Galena, sphalerite, pyrite, magnetite and pyrrhotite. Two compartment shaft. Country rock is devitrified rhyolite of Castine formation.

Deer Isle Quadrangle

The entire island of Deer Isle rests on a base of pink granite famous for its unusual coloring and superior quality. Stonington, on Deer Isle, has cut building blocks of this beautiful stone for many of the best known buildings throughout the country.

Two hundred Stonington men worked for two years to prepare the stone from here for New York's Triborough Bridge.

On each side of the causeway between Little Deer Isle and Big Deer Isle are large pieces of serpentine, some pure and some mixed with granite. This soft, hardness 4, metamorphic material is sometimes polished and used for interior work.

The road from the highway to the quarries is strewn with pieces of a low grade rhyolite, the very acid, lava form of granite.

On Deer Isle, in part, and on Little Deer Isle occur asbestos fibres in the fracture of the country rock, greenstone.

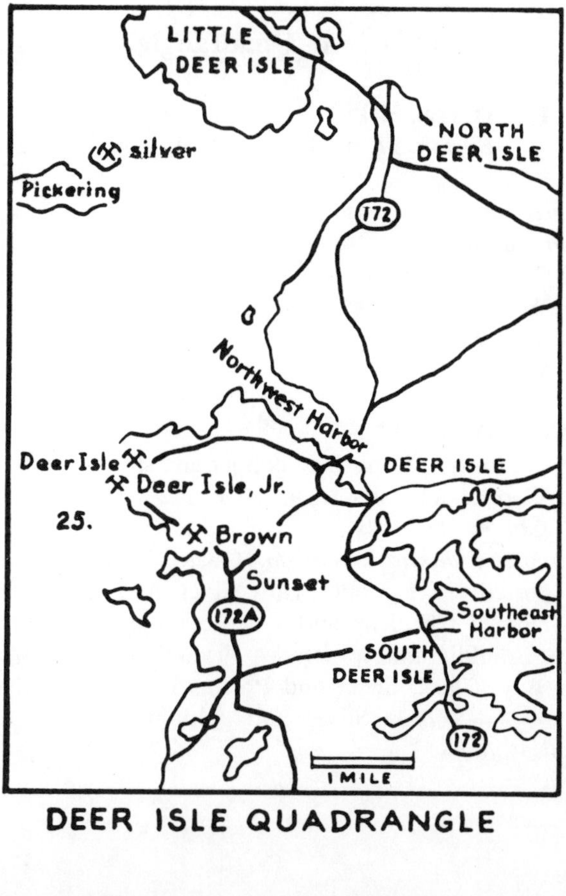

DEER ISLE QUADRANGLE

DEER ISLE TOWNSHIP

Map 25 . . . *Azure.* Reported: galena. Shaft 40 feet deep.

Belle of Deer Isle. Galena.

Brown. Silver.

Deer Isle. Sphalerite, pyrite, galena, chalcopyrite, native silver, magnetite, tetrahedrite (copper, zinc, lead, gold, silver and nickel). Shaft 260 feet deep. High grade ore reported as carrying 30% zinc, 16% Pb, 2½% copper and gold $2.50 a ton.

Deer Isle, Jr., Portland Deer Isle.

ISLE AU HAUT TOWNSHIP

Map 26 . . . *Burnt Island.* Reported: iron.

Gross, Isle au Haut. Copper and silver.

Ellsworth Quadrangle

ELLSWORTH TOWNSHIP

Map 27 . . . *Boston Silver.* Reported: argentiferous galena (lead and silver).

Brimmer. Silver.

Doneton. Silver.

Dunham. Silver.

Eastern Star. Gold.

Ellsworth. Pyrite.

Fulton, Gerry. Silver.

FRANKLIN TOWNSHIP

Map 28 . . . *Butler.* Reported: copper and silver. *Clark and Clark Extension.* Copper and silver.

Egypt. Copper and silver.

Franklin and Franklin Extension. Owned by West Franklin Mining Co. Reported: gray copper, ruby silver, native silver and galena (copper). Shaft 35 feet deep and a two compartment shaft 115 feet deep. The "Franklin lode" includes Franklin, Franklin Extension, Frenchman Bay, Clark, Swan and Portland Mining Exchange.

Silver Brook (Swan). Silver and gold in quartz.

State of Maine.

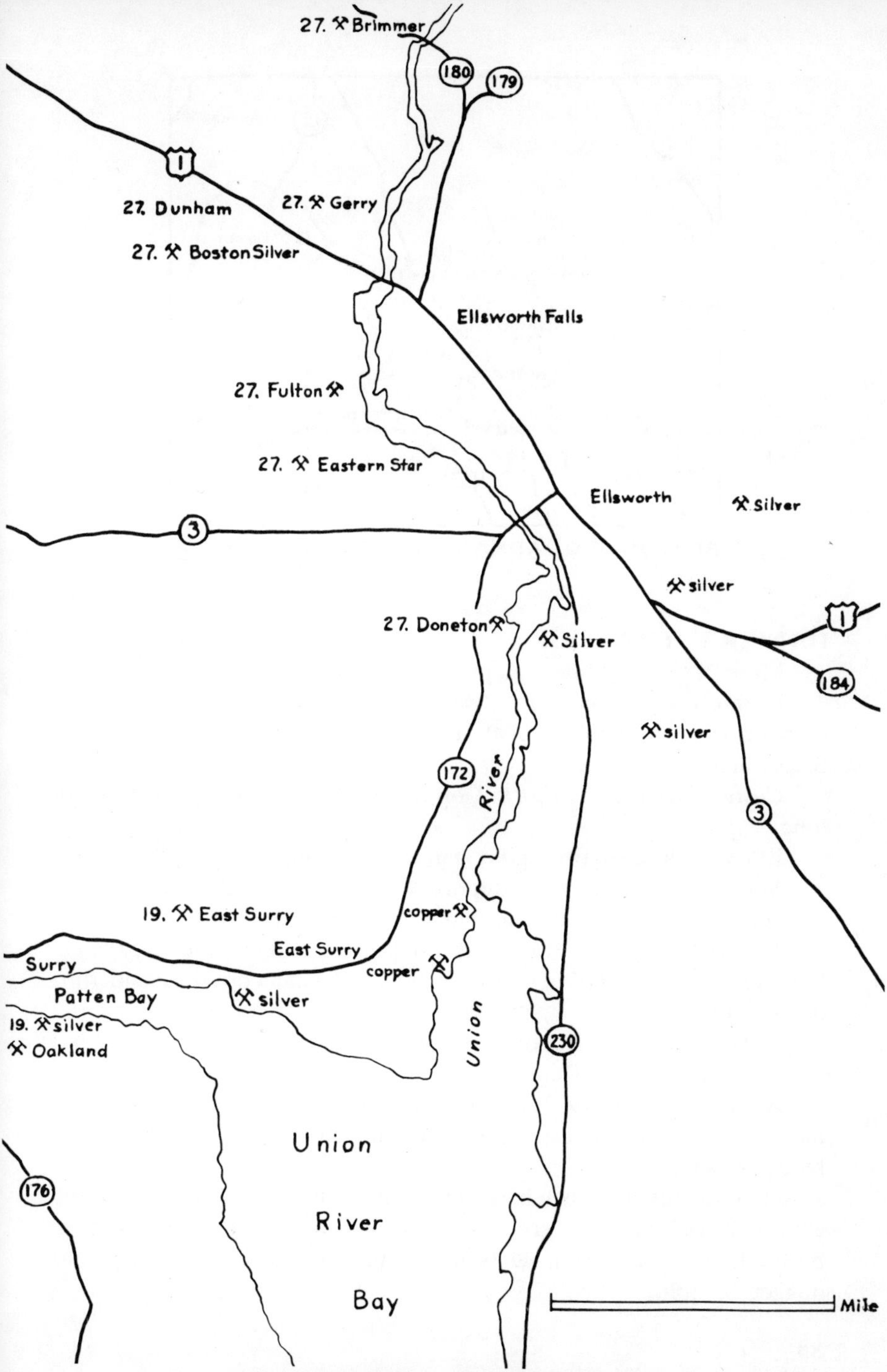

ELLSWORTH QUADRANGLE

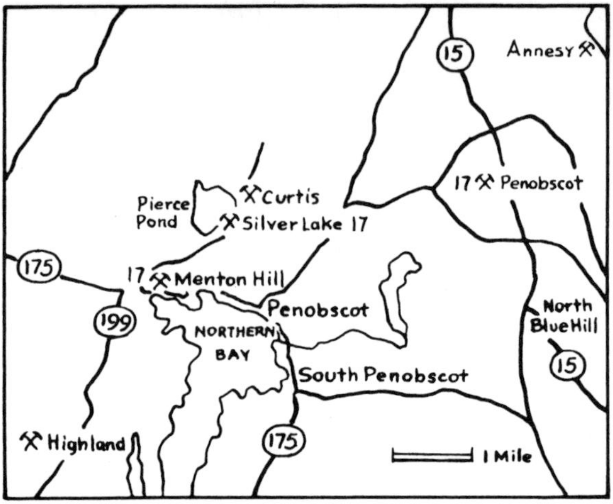

BLUE HILL QUADRANGLE - PENOBSCOT AREA

HANCOCK TOWNSHIP

Map 29 . . . *Cline.* Reported: silver and copper.

Custer (Copperopolis). Pyrite, chalcopyrite and rhodochrosite (copper, silver and gold). Shaft 110 feet deep. Highest grade 40 feet S. of shaft.

Custer Extension (Elliot). Copper and silver. Vein over 250 feet long.

Elliot. Chalcopyrite (gold, silver and copper). 110-foot shaft.

Map 29 . . . *Hagen.* Reported: sulphurets (silver and gold).

Hancock. Reported: sulphurets (silver and gold).

Hancock County. Reported: sulphurets (silver and gold).

Huronian — adjoins Custer. Reported: chalcopyrites and rhodochrosite.

Robert Emmet. Reported: copper, lead and silver.

Western Union. Reported: copper and silver.

Along Highway No. 1, at the end of the bridge going over Frenchman Bay to Sullivan, are mines near the tidewater with good green feldspar (amazonite).

Here in Ellsworth is where young Herbert Thurston picked up the extremely heavy, shiny green stone which, after being tripped over by his family for sixteen years, was discovered to be a seven pound nugget of gold.

The American Metals is prospecting in Ellsworth, locale of the "Ellsworth schist," for deposits of copper, lead and zinc. They have testing laboratories in trailers. They used helicopters for their reconnaissance.

Lead Mountain Quadrangle

DEBLOIS TOWNSHIP

Map 30 . . . *Leighton.* Reported: vein with silver.

Orland Quadrangle

BUCKSPORT TOWNSHIP

Map 20 . . . Silver.

Swans Island Quadrangle

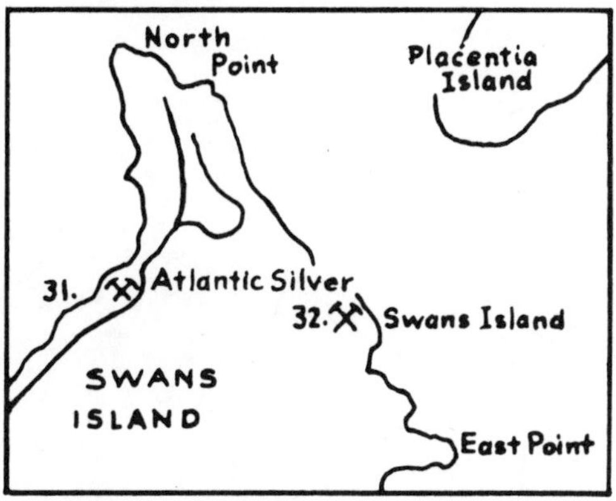

SWANS ISLAND QUADRANGLE

LONG ISLAND TOWNSHIP

Map 31 . . . *Atlantic Silver.* Reported: silver.
Black Island. Hydrated oxide of iron.

TRENTON TOWNSHIP

May 32 . . . *Sawyer.* Reported: silver.

SWANS ISLAND TOWNSHIP

Map 32 . . . *Swans Island.* Reported: gold.

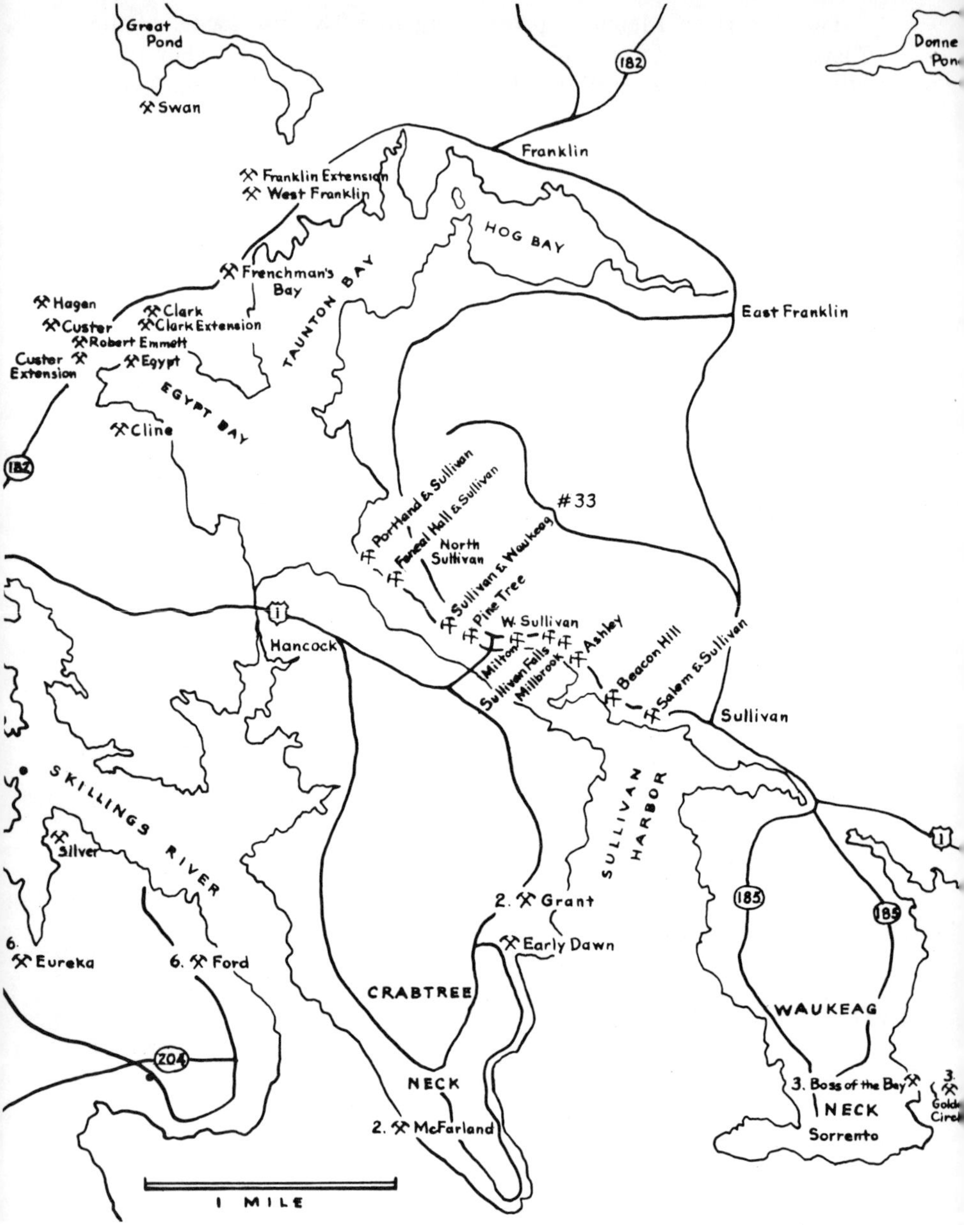

TUNK LAKE QUADRANGLE

Tunk Lake Quadrangle

Sullivan, in this quadrangle, has about everything — most of the metals and plenty of minerals besides.

Catherine Hill, in T10 S.D. has the world's largest deposit of molybdenum "white gold" along the shores of Tunk Pond. This metal which makes the superior molybdenum steel needed for airplanes and space ships would cost more to recover than it would sell for in the market. Another — "BUT"! Just wait, though, until Quoddy Dam is completed. The cheap power it will provide will change the above picture, and many other pictures in Maine.

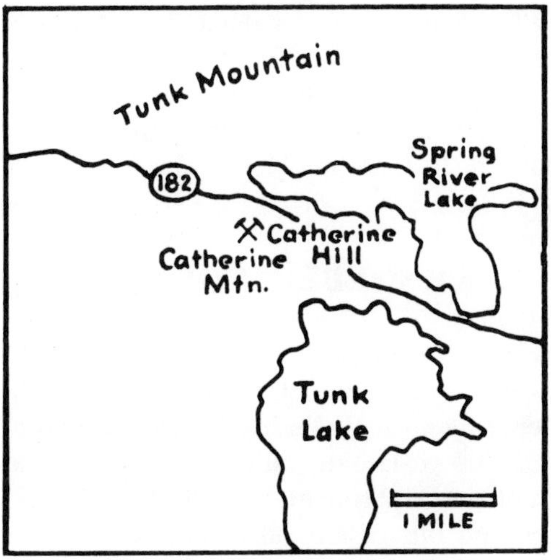

TUNK LAKE QUADRANGLE
Catherine Hill

SULLIVAN TOWNSHIP

Map 33 . . . *Ashley.* Reported: pyrite and galena. Located on the Sullivan vein. Vertical shaft 125 feet deep.

Beacon Hill. Pyrite and galena (lead, zinc, copper and silver) in a quartz vein.

Faneuil Hall and Sullivan. Gold and silver in a 20-foot wide vein. Shaft 165 feet deep.

Millbrook. Silver and gold in quartz stringers. Shaft 105 feet deep. Outcrop in Millbrook. Located on Sullivan vein.

Milton. Pyrite, chalcopyrite and native silver (lead, silver and copper). Two shafts, 160 and 180 feet deep.

Pine Tree. Galena. 126-foot shaft on the Sullivan vein.

Portland and Sullivan. Silver in a well mineralized quartzite vein on the Sullivan vein.

Salem and Sullivan. Pyrite, chalcopyrite, argentiferous galena, and native silver (lead, copper and silver). In a mineralized vein.

Sullivan (consolidated with Waukeag). Pyrite, chalcopyrite, stephanite, ruby silver, native silver, argentiferous galena, arsenopyrite (copper, lead, silver, zinc, gold-traces). Shaft 290 feet. Outcrops below high tide. On Sullivan vein. Country rock is the Ellsworth schist.

Sullivan Falls. Argentiferous galena (lead and silver). 680 feet of vein on the Sullivan vein.

Waukeag (consolidated with Sullivan). Same as Sullivan. In 1887, a cofferdam was built and a 14-stamp mill and smelting furnace operated. Produced 25 tons of silver — lead bullion until the mine was closed in 1884.

KENNEBEC COUNTY

On the shores of the Sandy River, New Sharon, in the Norridgewock Quadrangle in this county, Dr. Dabney Caldwell, instructor in the Geology Department at Wellesley College, found a four-foot red cedar log over 35,000 years old. This touched off an extensive survey by our State Geology Department for more specimens of wood, tree and plant pollen and bits of animal bone and teeth as guides to the climate, flora and fauna in prehistoric times. During the search, Dr. Caldwell found evidence of Maine's most recent glacier and below it an earlier one. He gave me slivers of fossilized wood found between the two glaciers. These have been carbon 14 tested and are about 40,000 years old. I touch them with awe and reverence.

Not far from Thirty-Mile River, near Manchester, in the Augusta Quadrangle, is a big boulder in a church wall imprinted with what legend claims are the footprints of the devil. Scientists state that the sixteen-inch-long human footprint of Maine's "Abominable Stoneman," followed by the much smaller one, undoubtedly of his woman he was dragging by the hair to his cave, are authentic petroglyphs carved long ago by remote Indians for ceremonial purposes. In Indian chronicles written over three hundred and fifty years ago these and the

KENNEBEC COUNTY

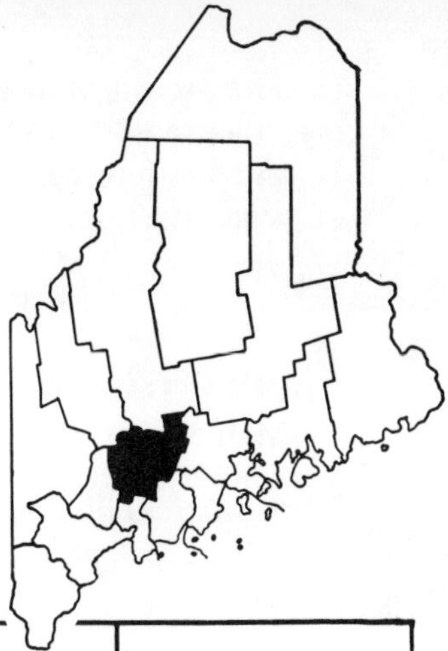

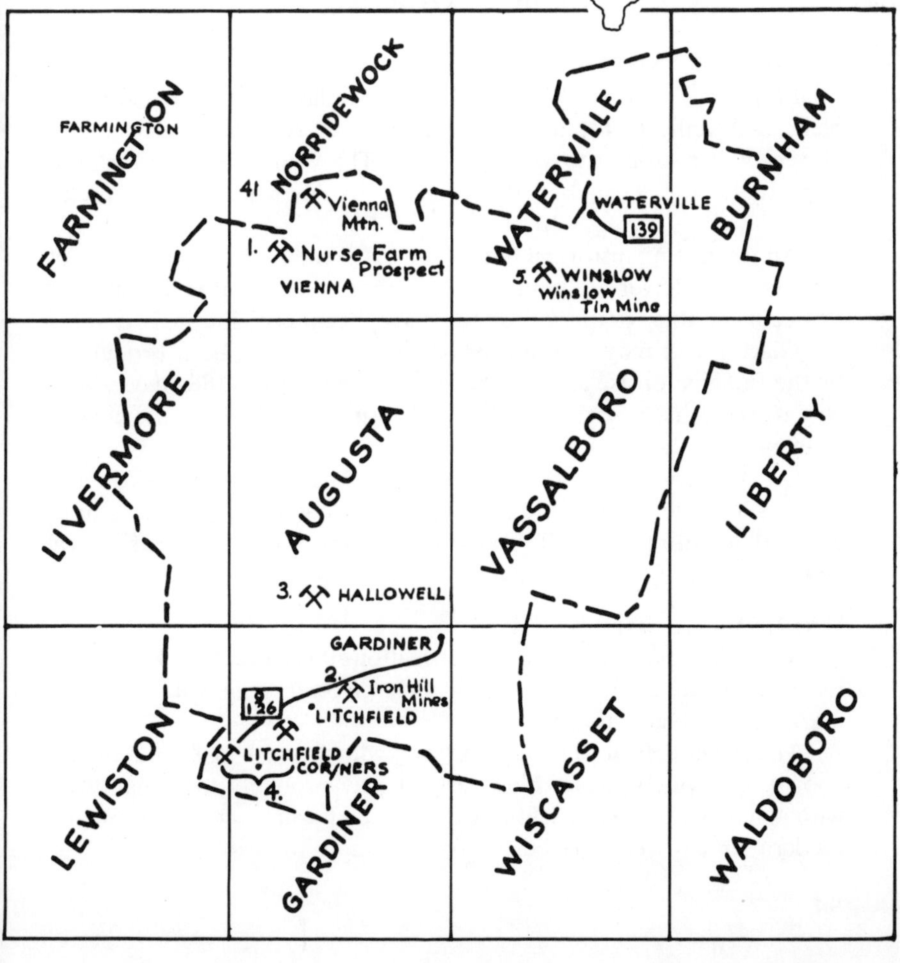

Dover-Foxcroft/Milo petroglyphs are referred to as "the stone carvings 'they' made." Who were the "they," the Red Paint Indians maybe, and how long ago did they inhabit Maine, perhaps many thousands of years ago?

Farmington Quadrangle

VIENNA TOWNSHIP

Map 1 . . . *Nurse Farm Prospect.* Owned by Roger Nurse, Mt. Vernon. Reported: muscovite and feldspar.

Vienna Mountain. Feldspar.

Gardiner Quadrangle

GARDINER TOWNSHIP

Map 2 . . . *Iron Hill.* Reported: pyrrhotite (nickel), sulphide blebs and veins in biotite granite gneiss. It is hoped the iron ore present will be found to yield sulphur. The vein runs from the iron mine under the Kennebec River into Pittston, where also is found silver.

An interesting historical note arises from the fact that the shores of the Kennebec River are peppered with flint. Since flint is thought not to occur in this vicinity, it must have been brought in.

Glacial drift may account for it. Or, it may have been brought in by the Indians for use as firelighters and arrow tips as they made their way to the coast to fish. We find them in shell heaps on the coastal islands, and shores, too.

Or, they may be gun flints and firelighters from when our Colonial soldiers and the French and later the Red Coats were pot-shotting at each other. Flint is usually found around the sites of our early forts and stockades.

HALLOWELL TOWNSHIP

Map 3 . . . Here is the source of Hallowell granite, long known for its exceptional quality and beauty. It is clear gray with pinpoint markings of black tourmaline.

The granite from various parts of Maine used to be nationally known and widely used. Now less so, due probably to competition with vinyl, glass, rayon, skimmed milk and whatever. Besides, the modern creed seems to be, "Don't build it to last."

A number of years ago a New York contractor dug up an old street and was amazed at the fine state of preservation of the granite after forty years use. When he learned that the stone had come from Oak Hill Quarries in Maine, the result was an order to ship forty-five thousand blocks a week to pave a forty million dollar boulevard in New Jersey. A Boston editor commented, "How many more dead quarries are there in New England waiting to be rediscovered by the smart and wide-awake gentlemen from the other side of Connecticut?"

Former State Geologist John Rand lists 161 quarries in Maine which have supplied granite for over 350 buildings in various states of the nation.

LITCHFIELD TOWNSHIP

Map 4 . . . *Two Pastures.* Reported: sodalite, cancrinite, litchfieldite, very black mica and crystals of brown zircon, some of which fluoresce.

Sodalite is deep blue and opaque like lapis lazuli, an entirely different mineral family but colored blue by the sodium common to both. It cuts into a handsome gemstone and makes a colorful specimen when streaked with the bright yellow cancrinite.

Litchfield is a widely known locality for sodalite which occurs in two pastures about a mile from Gardiner near Litchfield Village.

One boulder-filled pasture is on the right one mile north of Gardiner, before you get to the church corner, Tacoma Inn and the first Litchfield Village.

The second is beyond the village, a mile east of Litchfield four corners on Highway No. 126 where the road makes a sharp angle with the main road. After the road goes uphill in an easterly direction and makes a hairpin turn to the left, the pasture is to be seen on the right, or north, side of the road. Numerous boulders and slabs are in the fields and the pasture across the road from Will Woodbury's house and in the open pasture near the road before you reach the Woodbury place.

Waterville Quadrangle

WINSLOW TOWNSHIP

Map 5 . . . *Winslow Tin Mine.* Reported: cassiterite (tin) in a shaft 100 feet deep and small white beryl crystals.

This, the only tin mine in the U. S., is located on a hillside on the Augusta-Waterville Road.

The mine was operated during the Civil War but has been closed since because it is thought to be unprofitable to extract the ore. Just as India has scooped Maine with manganese, I suppose Bolivia does the same with tin. Maybe someday, when theirs runs low, Maine may come into its own!

KNOX COUNTY

Here is where everyone had high hopes for the Union Pyrrhotite Prospect to be operating as the only nickel mine in the U. S. World economy makes the actual mining unprofitable, but the nickel will always be there should we need it.

For over a hundred years this Penobscot Bay area has been a nationally important source of limestone. The Martin Marietta Cement Co. of Thomaston at present conducts the largest mineral enterprise in the State. Some of these days if you are sidewalk supervising the construction of a building many hundreds of miles away from Maine you will probably see cement being poured which may very well have come from the huge cement plant in Thomaston.

When Jack Rand was State Geologist, I asked him what he considered the best mining bet in Maine, expecting him to say either "Oxford County's pegmatites with their fantastic variety of rarities and gemstock" or "Hancock County, so richly endowed with metals."

But, no. It was Knox County limestone for cement or agricultural purposes. "This is the easiest, cheapest and surest in the State. There is a guaranteed, captive market right here in Maine. Using the most efficient, up-to-date methods, a good profit is assured with a relatively small capital outlay."

Attention Wall Street! Me, I'd rather go after gold and tourmalines and aquamarines — and lose my shirt.

Belfast Quadrangle

APPLETON TOWNSHIP

Map 1 . . . *Appleton.* Reported: chalcopyrite, pyrrhotite, tourmaline, garnet and copper.

KNOX COUNTY

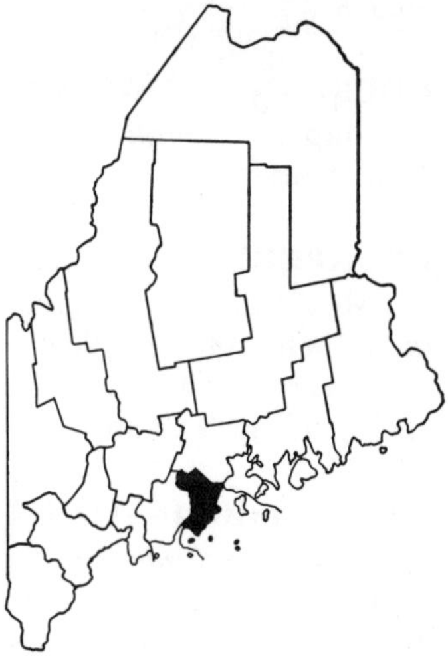

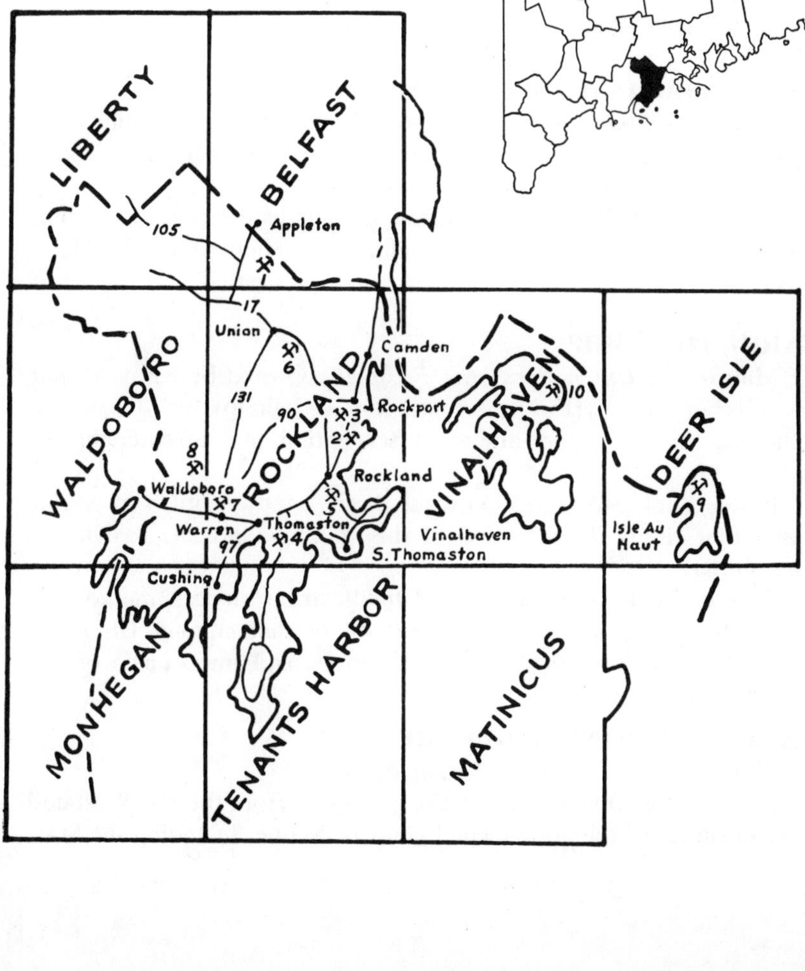
LIBERTY
BELFAST
Appleton
105
17
Union
6
Camden
WALDOBORO
ROCKLAND
131
90
3
Rockport
2
8
Waldoboro
7
Rockland
5
Warren
Thomaston
97
4
S. Thomaston
Cushing
VINALHAVEN
Vinalhaven
10
DEER ISLE
9
Isle Au Haut
MONHEGAN
TENANTS HARBOR
MATINICUS

Rockland Quadrangle

Here is the famous limestone and nickel area. And in this quadrangle, too, is the State Prison Farm, a highly metal and mineralized area.

ROCKLAND TOWNSHIP

Map 2 . . . *Pratt, Rockland City.* Reported: argentiferous galena in a quartz vein.

ROCKPORT TOWNSHIP

Map 3 . . . *Porterfield.* Reported: copper, silver, gold and nickel in a shaft 100 feet deep.

Riverside. Reported: arsenopyrite (copper, nickel, lead, silver and gold) in a shaft on a ridge 200 feet above water level.

SOUTH THOMASTON TOWNSHIP

Map 4 . . . *Limestone area.* Reported: limestone, crystalline limestone, marble, fans of fine white tremolite crystals, beryllium, sphalerite (zinc) and various other metals.

Map 5 . . . *Owl's Head.* Reported: gold and silver. In 1939, some gold and silver were found in several shafts. There is a rich vein of silver.

UNION TOWNSHIP

Map 6 . . . *Union Pyrrhotite Prospect.* Owned by Knox Mining Co. Reported: pyrrhotite, pentlandite, chalcopyrite, magnetite, sphalerite (nickel, copper and zinc), serpentine, talc, arsenic, chlorite and marble.

It was John Houston's doctoral thesis on pyrrhotites in Maine which first called the attention of this mining syndicate to Union's possibilities.

This is the prospect I selected to illustrate modern methods of mining for metal in Maine, Chapter 4. In that chapter BE PREPARED I have discussed the specifications and importance of this prospect in full detail.

WARREN-CUSHING TOWNSHIP

Map 7 . . . *The State Prison Farm.*

In 1956, the Dow Chemical Co. geologists from the Far West and Canada surveyed the State-owned lands of Maine, including the State

Prison Farm. So far, they have found thirty dikes of spodumene containing the lightest of metals, lithium, used among other things, to trigger the hydrogen bomb. This occurs in a hundred and eighty acre section here.

In 1958, the State of Maine Geology Department explored this highly mineralized area. The results of this preliminary exploration, as of the others they have done in the State, are available to a prospective operator.

Waldoboro Quadrangle

WARREN TOWNSHIP

Map 8 . . . *Starrett Mine and Starrett Prospect.* Owned by Hollis G. Starrett, Warren. Reported: feldspar (microcline), muscovite, quartz, spodumene, beryllium (the Maine "only" — from white cesium beryl), columbite, tantalite (used for light filaments) and a host of other minerals and metals.

Mr. Starrett's thirty-six acres on the West Station road in Warren contain large crystals of lithium-bearing spodumene, probably the largest bed of this rare space metal in the east.

Deer Isle Quadrangle

ISLE AU HAUT TOWNSHIP

Map 9 . . . *Burnt Island.* Reported: iron.

Gross, Isle au Haut. Copper and silver.

Vinalhaven Quadrangle

VINALHAVEN TOWNSHIP

Map 10 . . . *Island City.* Copper and silver in three veins 8 feet wide at 19-foot level.

LINCOLN COUNTY

Boothbay Quadrangle

Geologically the Boothbay Region is notable for its clear lines of demarcation at the contact of granite with pegmatite. But, although the pegmatite zone which becomes so rich in Oxford County starts here this region yields comparatively little of interest mineralogically.

LINCOLN COUNTY

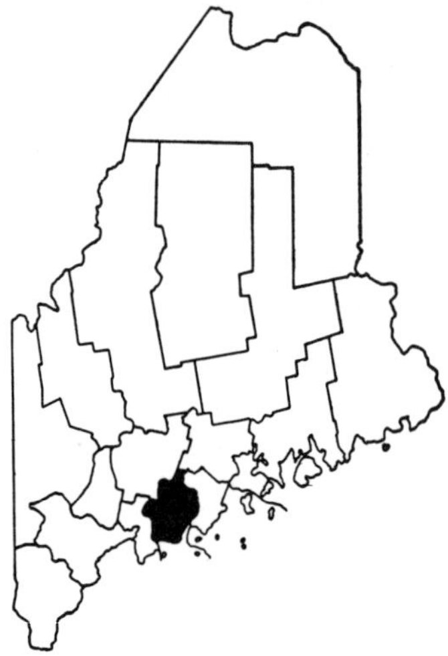

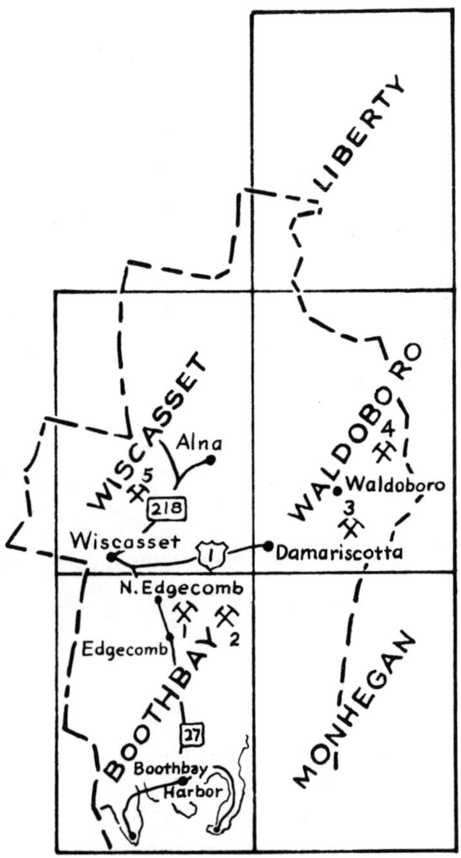

EDGECOMB TOWNSHIP

Map 1 . . . *Poole Prospect.* Owned by a Mr. White, West Bath. Reported: feldspar, muscovite, quartz and beryl.

Mr. White, a lapidary, has cut some very fine smoky quartz gems and some aquamarine from material he obtained here.

Map 2 . . . *Williams Mine.* Owned by Frank Williams, Edgecomb. Reported: feldspar, muscovite and quartz.

This Mt. Hunger area has several quarries, one large and deep filled with water. Near it is abandoned mining machinery.

Waldoboro Quadrangle

DAMARISCOTTA AND DAMARISCOTTA TOWNSHIPS

Map 3 . . . Beryl has been reported on the shores up the Damariscotta River.

This is the area of the world's largest Indian oyster shell heaps, kitchen middens, largest heaps, whaleback being one hundred and twenty-five feet wide and thirty-one feet deep, and largest oysters, one taken from the upper strata of whaleback measuring one foot long and eight inches wide. It was originally sixteen inches long and would have filled a pint measure. These middens go back further than the historical records of our own Indians to the Red Paint People, who are thought to have antidated Christ.

While we are digressing from minerals, it was here, about one hundred years ago, that a lad from the village watched pirates burying a treasure chest. The contents launched him on what became a fabulous Wall Street career. Harold Castner, of Damariscotta, knows his name but loyally refuses to divulge it. He said we all know of the famous man. Didn't I hear that the John Jacob Astor fortune was built on a nest egg of buried pirate treasure?

After a jaunt to Damariscotta Mills to watch the alewives leap the falls in the early spring, we might swing around to Alna, Wiscasset quadrangle, to try to locate the highly mineralized area Bob Doyle, State Geologist, discovered. He showed me a bottle of beautiful crystals of molybdenite, the rare metal known as "white gold."

WALDOBORO TOWNSHIP

Map 4 . . . *Benner Prospect.* Owned by Merton Benner, Waldoboro. Reported: feldspar, muscovite and quartz.

Wiscasset Quadrangle

WISCASSET TOWNSHIP

Map 5 . . . *LaPointe Prospect.* Owned by Paul LaPointe, Wiscasset. Reported: biotite, feldspar and quartz.

ALNA TOWNSHIP

Map 5 . . . Molybdenite reported.

OXFORD COUNTY

The rhythm of my tribute below to Oxford County, whose pegmatites are the most minerally varied on earth, is borrowed from *Hiawatha* by Maine's own Henry Wadsworth Longfellow, who now sits in stone in Longfellow Square in the city of his birthplace, Portland.

To The Oxford Hills

You shall hear how Oxford County
Leads the world in mineral riches.
How her pegmatites are teeming
With every type of gem and mineral.
How her rare earths, oddballs, far-outs
Speed the rockets into space.

Her pockets in the rocky mine face
Spill out crystals bright in hue,
Green and rose and multicolored,
Purple, blue and tawny yellow.
Once in many moons a great find!
Now, the largest beryl crystal
Rising house high to the mine top.

Amethyst of royal purple, tourmaline of ruby hue,
Emerald of the finest water;
Each a royal crown enhances.
Museums the whole world over
Show her large and flawless crystals.

In her rushing rivers gold flakes,
Citrines in her water falls.
Famed in fact and told in legend,
Hear her sparkling, startling story.

OXFORD COUNTY

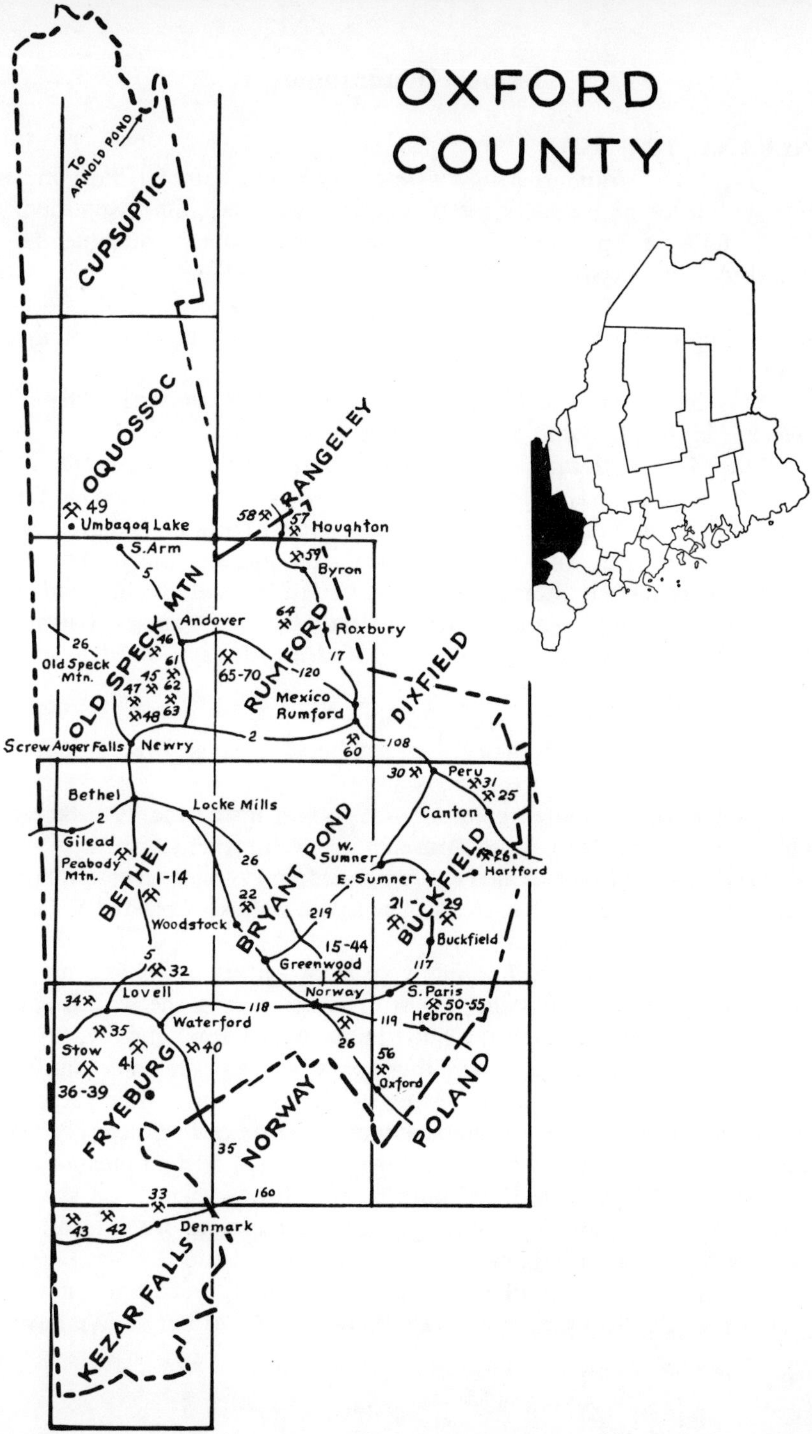

Bethel Quadrangle

ALBANY TOWNSHIP

Map 1 . . . *Bumpus Mine.* Owned by Ralph Kimball, Portland, who bulldozes periodically for rockhounds who pay him. Reported: beryl, feldspar (plagioclase), muscovite, rose quartz, pitchblende, tantalite and beryllium.

Map 2 . . . *Donahue Prospect.* Owned by Ralph Kimball, Portland. Reported: beryl, feldspar, muscovite, quartz and black tourmaline.

Map 3 . . . *Pingree Prospect.* Owned by C. P. Pingree, Albany. Reported: beryl, columbite and feldspar.

Map 4 . . . *Rattlesnake Mountain Area.* Reported: beryl.

Albany, MAINE. Here at a crossroads, deep in the woods far from human habitation, is the famous signpost so often pictured on postcards. It lists with mileages, local and foreign, the dozens of Maine towns named after faraway places. These come to mind, and at that I may be skipping a few: Albany, Bath, China, Denmark, Gilead, Egypt, Hebron, Paris, Peru, Poland, Madrid, Mexico, Sweden, and Vienna.

And the star of the area, Bumpus Mine.

From here came the world's largest beryl crystals, Nos. 1, 2 and 3, each larger than the previous.

As I have mentioned before, No. 2 is just inside the 79th Street entrance of The New York Museum of Natural History.

And No. 3, also described earlier, ended up in small pieces on box cars. You will recall that it was Dana Douglas who blasted it out, then ordered it smashed.

When I asked Dana if so many years of getting out quantities of rose quartz (Bumpus has the largest concentration in the world) and beryl have about depleted the mine he answered emphatically, "No! There is three times as much there now as we ever took out."

Dana also told me this interesting story. As he was walking across a clearing in Albany, he kicked back some moss, revealing clear crystal quartz. When he rolled back the carpet of moss, he had cleared an almost room-sized window of pure crystal through which he saw a large, perfect aquamarine crystal. Taking careful note of the location he planned to return later with tools to get it out.

He knows the area like the back of his hand, yet he has never since been able to locate the quartz-bound crystal of beryl. We have all had happen to us what Dana said, "If you don't take what nature sends you when she gives it, she will take it back again."

He said, too, "Whatever man can dream up of beauty or surprise, nature can top." Bumpus Mine has proved that again and again.

Many years ago, even before the first of the record-breaking beryl crystals, Bumpus had an international reputation for its beauty. A group of scientists from all over the world convened at Harry Bumpus' mine.

At that time huge crystals of beryl; green (emerald), golden and aquamarine pushed out of a whole cliff of pure rose quartz like log jam, the whole reflected in the pool of water in the bottom of the mine.

All agreed it was the finest thing of the kind they had ever seen. The tears streamed down the cheeks of an old German professor as he said, "Dod is the most BEAUDIFUL sight I have ever seen!"

BATCHELDERS GRANT TOWNSHIP

Map 5 . . . *Peabody Mountain Quarry.* Owned by Floyd Mason, West Bethel. Reported: apatite, feldspar, muscovite, quartz, and tourmaline.

GILEAD TOWNSHIP

Map 6 . . . *Peaked Hill and Prospect.* Owned by Mrs. H. E. Mason or Dr. W. B. Twadale, Bethel. Reported: biotite, feldspar, garnet, muscovite, quartz, and gem beryl.

Wheeler Mines. Owned by Roger and Chester Wheeler, Bethel. Reported: beryl, feldspar (cleavelandite), and muscovite.

MASON TOWNSHIP

Map 7 . . . *Anderson Mine.* Owned by Lawrence Anderson, East Stoneham. Reported: muscovite.

STONEHAM TOWNSHIP

In 1900 a Stoneham miner found a beautiful gem crystal which he set out on his front porch because the color shone so prettily in the sunlight. A passing traveller bought it for a few dollars as a specimen. Shortly after, a picture of the Hapsburg crown jewels showed this same gem in the royal collection. Investigation showed that Prince Henry of Prussia had bought the Stoneham specimen, thought to be an amethyst or possibly an aquamarine, from a New York gem dealer for $30,000.00.

Maine gems had a habit of finding their way onto European crowned heads and into royal museums as I'll tell more about under Mt. Mica.

Map 8 . . . *Aldrich.* Owned by Rupert Aldrich, South Paris. Reported: beryl, feldspar, muscovite, pyrite and serpentine.

Andrews Ledge. Owned by Mrs. W. N. Swett, Norway. Reported: beryl and muscovite.

Butters Mountain Area. Reported: golden beryl and garnet.

Durgin Mountain Area. Reported: beryl.

Foster Hill Prospect. Owned by James Merrill, East Stoneham. Reported: feldspar and muscovite.

Map 9 . . . *Lord Hill Mine.* Supervisor, White Mountain National Forest; U. S. Bureau of Land Management, Washington, D. C. Reported: beryllium minerals, columbite, amethyst, and gem topaz, golden beryl, purple fluorite, green and white apatite, pink and blue feldspar, rose quartz, smoky quartz crystals, chrysoberyl, phenucite, ocrite crystals and the largest gem crystals of beryllonite, as well as the finest sheet mica.

This potash-feldspar ledge is one of the chief sources of topaz in Maine. It was first found by Berry in the early eighties in large massive crystals. An outstanding topaz find was made here. One of the finest gem aquamarines was also found here.

Map 10 . . . *Melrose.* Owned by Mrs. Enid Melrose, West Stoneham. Reported: bertrandite, aquamarine beryl and beryllonite.

Map 11 . . . *Styles Mountain Prospect.* Owned by W. Phillips Cole, Hollis Center. Reported: beryl.

Map 12 . . . *Willis Warren Quarry.* Owned by Willis Warren, Stoneham and the late Stanley I. Perham, West Paris. Reported: apatite, beryl, feldspar, garnet, muscovite and tourmaline.

Bryant Pond Quadrangle

ALBANY TOWNSHIP

Map 13 . . . *General Electric Glass Quartz Prospect.* Owned by the Town of Albany. Reported: beryl, columbite, feldspar, garnet and quartz.

Holt Prospect. Owned by Chester Holt, Albany. Reported: muscovite.

Johnson Mine (Guy Johnson Mine) Owned by A. I. Bean, Albany. Reported: beryl, feldspar, muscovite, quartz and black tourmaline.

ALBANY AREA

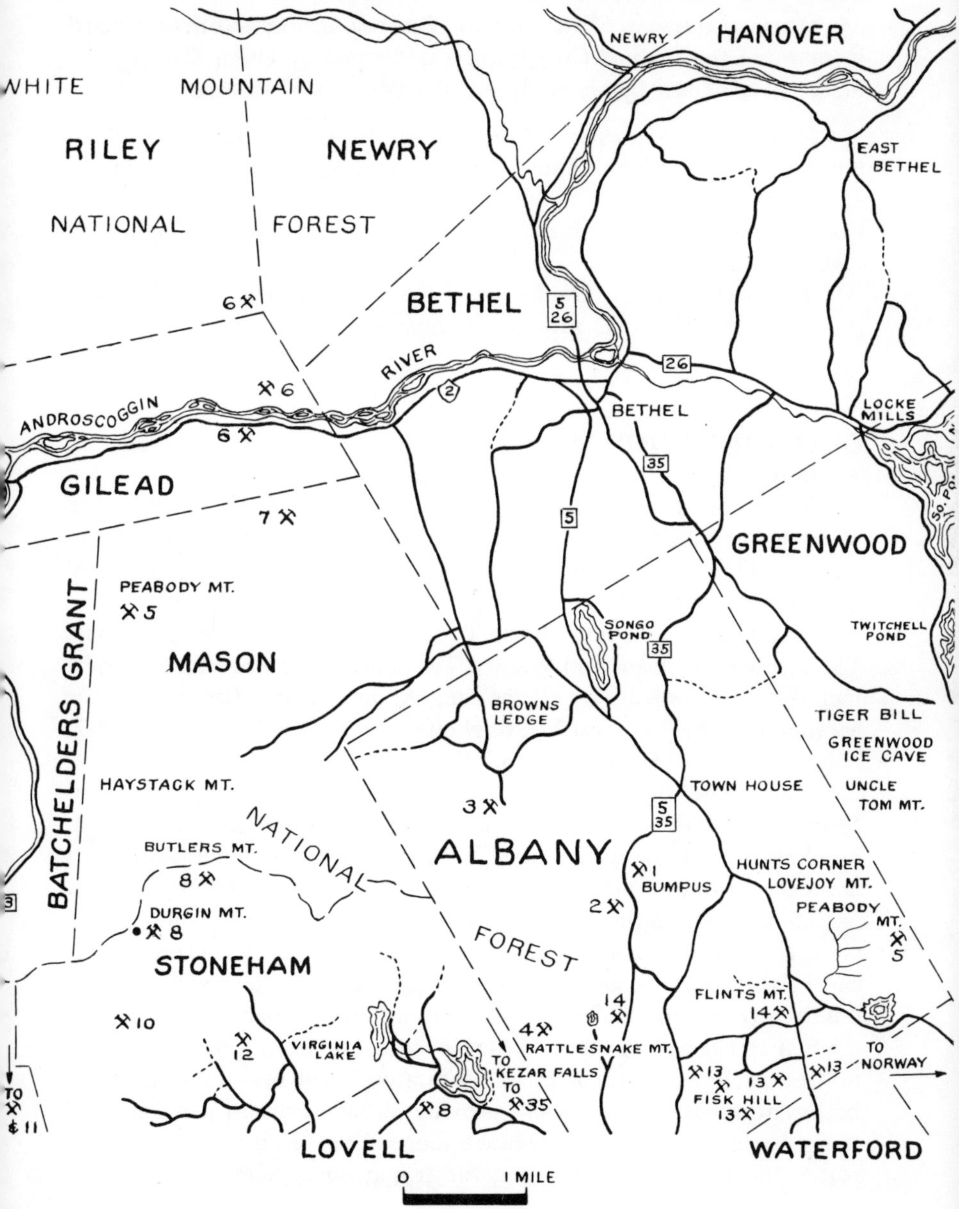
WHITE MOUNTAIN NATIONAL FOREST
RILEY
NEWRY
HANOVER
EAST BETHEL
BETHEL
ANDROSCOGGIN RIVER
GILEAD
LOCKE MILLS
GREENWOOD
PEABODY MT.
MASON
SONGO POND
TWITCHELL POND
BROWNS LEDGE
TIGER BILL
GREENWOOD ICE CAVE
BATCHELDERS GRANT
HAYSTACK MT.
TOWN HOUSE
UNCLE TOM MT.
NATIONAL FOREST
ALBANY
BUTLERS MT.
HUNTS CORNER
LOVEJOY MT.
BUMPUS
DURGIN MT.
PEABODY MT.
STONEHAM
FLINTS MT.
VIRGINIA LAKE
TO KEZAR FALLS
RATTLESNAKE MT.
TO NORWAY
FISK HILL
LOVELL
WATERFORD
0
1 MILE

Scribner Mine. Owned by Fred Scribner, Bryant Pond. Reported: apatite, beryl, biotite, feldspar, garnet, magnetite, muscovite, rose quartz and black tourmaline.

Stearns Prospect (Hornet Mine). By the way, hornets are a hazard in many of the Oxford County mines. Owned by Hugh Stearns, Albany. Reported: beryl, feldspar, muscovite and quartz.

Map 14 . . . *Wardwell Mine.* Owned by Roy Wardwell, Albany, and leased to The North Carolina mica interests described in Chapter 4, BE PREPARED, to illustrate a typical pegmatite mining operation. Reported: beryl, biotite mica, muscovite mica, feldspar and rose quartz.

Wentworth Mine. Owned by Ernest Wentworth, Albany. Reported: apatite, beryl, muscovite, quartz and pyrite.

GREENWOOD TOWNSHIP

Map 15 . . . *Greenwood Ice Caves.*

Several hundred years ago, landslides formed large caverns, one of which is called *The Cathedral.* This one is thirty feet wide and retains winter ice in its interior as late as July. May be the origin of our saying, "Maine has two seasons, Fourth of July and winter."

Greenwood is the hub of The T. C. Mining Company's operations. The late Stan Perham and Willard Titcomb prospected twenty mines, including Mt. Mica, Mt. Rubellite, Andy Brown, Slattery, Noyes Mountain, the G. E. and Westinghouse mines, BB No. 1 through 7, Harvard, Willie Heikkinen and Waisnen.

Here is the long and impressive list of what those mines contain: Mica: for electronic equipment and the production of condenser film. Spodumene: source of the metal, lithium, and lithium chlorides, used in air conditioning and in development of the hydrogen bomb. Cesium ore: five thousand times as powerful as T.N.T., used in photo-electric cells, transistors, standard infra-red signal units for the Army and in cancer therapy. Cesium 137 isotope has largely replaced the cobalt bomb in treatment of cancer. Beryllium; used to harden copper and other metal tools. Copper hardened with beryllium makes a spring that never loses its springiness. Tungsten: used in various types of bulbs and tubes. Columbium: provides a heat resistant metal needed in the building of jet planes. Black tourmaline: used in piezo electric plates for pressure gauges and electronic equipment.

GREENWOOD AREA

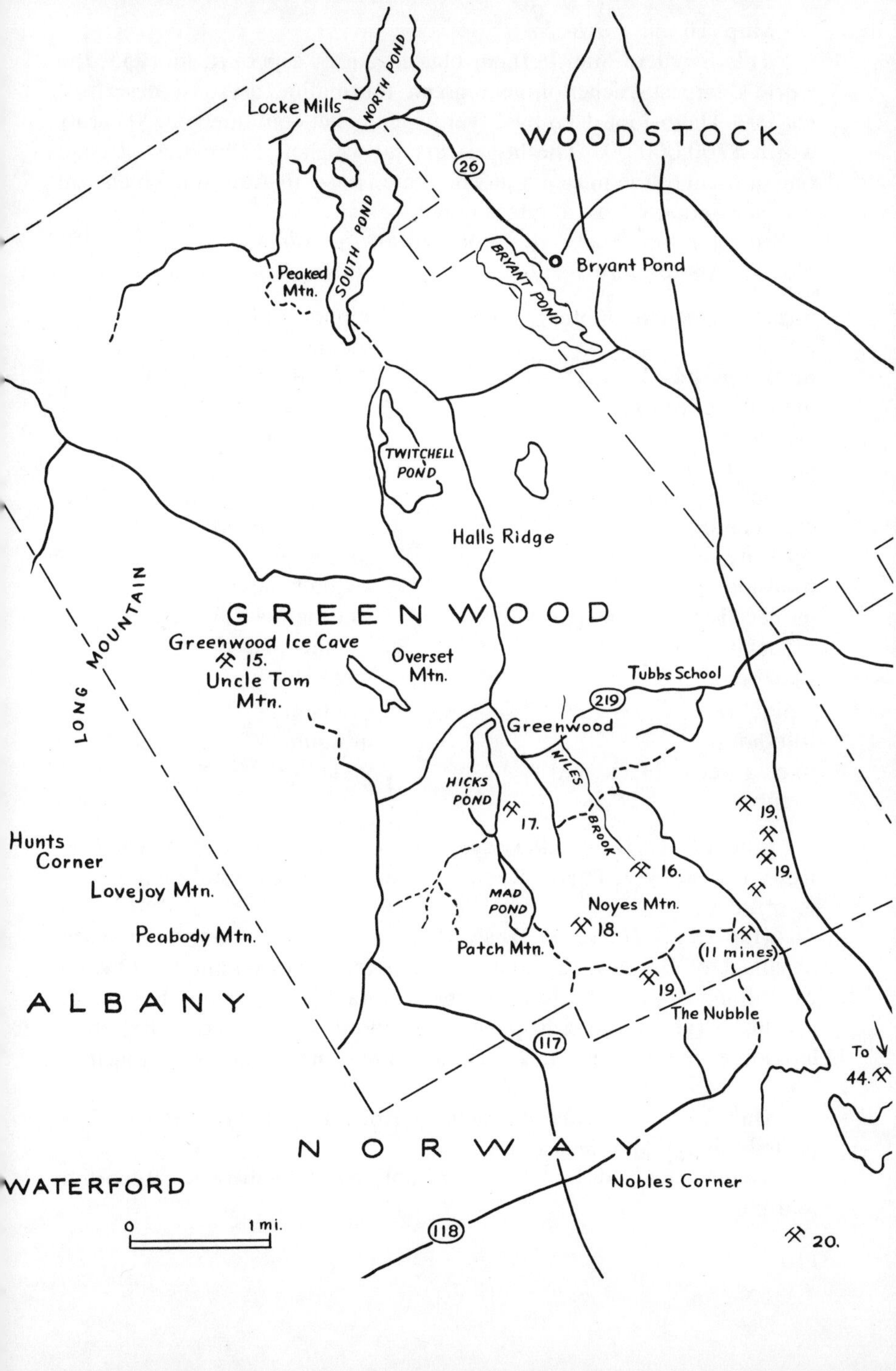

Map 16 . . . *BB No. 7.*

This is where Stan Perham blasted out of a pocket, in 1955, the world's largest cache of gem green tourmaline crystals, described earlier. The 6-foot deep by 2 feet wide pocket contained 6,000 carats worth $200,000.00. The largest crystal weighed 137 carats. I gave one of them to the museum at The State House in Augusta, which you can see mounted as a micromount.

Just for fun, Stan and I counted all the minerals present in BB No. 7. There were forty-eight — incredible, from one mine!

apatite (green and blue)
albite
arseno-pyrite
autunite crystals
cassiterite
chromium mica
columbite (with tantalum)
cleavelandite
cookeite
diopside, pink
garnet (red and mangana)
hiddenite (green)
kunzite
lepidolite
lillotholite
mica (green, black and ruby)
montmorillonite
moonstone
pollucite
purpurite
pitchblende
pyrite
quartz (milky, rose, smoky and banded)
scheelite
spodumene
tourmaline (green, pink, blue and black)
trypholite
urananite
uranium
vivianite

Map 17 . . . *Diamond Ledge (Yale or Yates Quarry).* Owned by Rolly Hayes, West Paris. Reported: quartz. The crystal quartz here is brilliant.

Map 18 . . . *Harvard (Noyes Mountain Quarry).* Four miles south of village of Greenwood, high on southwest end of mountain. Owned by William Bryant, Rumford, or the late Stanley I. Perham. Reported: apatite, beryl, biotite, feldspar (cleavelandite and microcline), garnet, lepidolite, muscovite, quartz crystal, spinel, black and green tourmaline.

Map 19 . . . *Hayes.* Owned by Rolly Hayes, West Paris. Reported: beryl and muscovite.

Heath. Owned by Nestor Tamminen, Greenwood. Reported: feldspar.

Heikkinen Quarry. Owned by the late Stanley I. Perham, West Paris. Reported: beryl, feldspar, muscovite, pyrite and quartz.

Immonen Ledge No. 2. Owned by H. Donald and Albert Penley, West Paris. Reported: fibrolite.

Mustonen Ledge. Owned by Roland Benson, West Paris. Reported: feldspar.

Ohtonen's Quarry. Owned by Anti Ohtonen, West Paris. Reported: apatite, beryllium minerals, feldspar and pyrite.

Tamminen Old Mine. Owned by Nestor Tamminen, West Paris or Bell Minerals Co., West Paris. Reported: beryl, feldspar (albite), pollucite, quartz crystal, spodumene, topaz and uraninite.

Tamminen-Waisenen Mine. Owned by the late Stanley I. Perham, West Paris.

Tiger Bill Mine. Owned by H. Donald and Albert Penley, West Paris. Reported: autunite, bertrandite, feldspar, herderite and spodumene.

Uncle Tom Mine (Emmons Quarry). Owned by H. Donald and Albert Penley, West Paris.

Waisanen Mine (The Nubble). Owned by the late Stanley I. Perham, West Paris. Reported: beryl, feldspar and muscovite.

MILTON PLANTATION TOWNSHIP

Map 20 . . . *Champion and Champion Consolidated and Oxford.* Reported: two shafts.

PARIS TOWNSHIP

Map 21 . . . *Hoopers Ledge (Twitchell Mine).* Owned by Bell Minerals Co., West Paris. Reported: beryl, chrysoberyl, feldspar and rose quartz.

Immonen No. 1. Owned by Emil Immonen, West Paris.

Perham Mine. Owned by Bell Minerals Co., West Paris. Reported: apatite, aquamarine beryl, biotite, feldspar, garnet, muscovite, pyrite, quartz and tourmaline.

WOODSTOCK TOWNSHIP

Map 22 . . . *Lone Star.* Reported: metals in 2 shafts.

Woodstock. Reported: copper, lead, silver and gold.

Buckfield Quadrangle

Autunite, important host of uranium, is found near Buckfield.

BUCKFIELD TOWNSHIP

Map 23 . . . *Bennett Quarry.* Owned by Mrs. Blanche Bennett. Reported: amblygonite, apatite, arsenopyrite, beryl, cassiterite, columbite, feldspar, garnet, lepidolite, muscovite, pollucite, quartz, rhodochrosite, spodumene, tantalite, topaz and tourmaline.

This is a typical soda-feldspar ledge pegmatite. In all about 45 minerals have been reported from here and the nearby Orchard Mine. Many are rare, or oddities, and many fluoresce.

The museum of the Sheffield School of Yale University has an excellent twinned chrysoberyl crystal from here.

Charlie Bragg said that Blanche Bennett had a large nugget of what she took to be rose quartz in her dooryard. Geologists from Ohio stopped by and recommended she take it indoors. It was a forty carat nodule of rare and very valuable gem pink cesium beryl.

Map 24 . . . *Bessey Quarry.* Owned by the late Stanley I. Perham, West Paris. Reported: arsenopyrite, beryl, feldspar, muscovite, smoky quartz with liquid inclusions and sphalerite.

Cummings Mine. Owned by Horace Cummings, Buckfield. Reported: quartz.

Dudley Prospect (Neville-Owl's Head). Owned by Alfred McKay, Norway. Reported: amblygonite, beryl and pollucite.

Fletcher Mine. Owned by estate of Howard Irish. Reported: beryl.

Irish Mine (Westinghouse Mine). Owned by estate of Howard Irish.

Orchard Mine. Owned by Mrs. Blanche Bennett. Reported: apatite, gem beryl, cassiterite, columbite, feldspar, rose quartz, spodumene, topaz and gem tourmaline.

CANTON TOWNSHIP

Map 25 . . . *Clark Mine (Wentworth Mica Mine).* Owned by Charles House, Rumford Center.

HARTFORD TOWNSHIP

Map 26 . . . *Ragged Jack Mountain.* Owned by Wesley Poland, Woodstock, or Clarence Lefoy, East Sumner. Reported: chrysoberyl (chatoyant — the rare and valuable gem "cat's-eye" chrysoberyl).

PARIS TOWNSHIP

Map 27 . . . *Bowker Mine.* Owned by estate of Howard Irish. Reported: rose quartz.

Map 28 . . . *Mt. Mica.* Owned by Plumbago Mining Co. Reported: tourmaline — green, watermelon, cucumber, pink, rose (rubellite), blue (indicolite), black (schorl) and white (achroite).

Mt. Mica! Magic name! Here it was that tourmaline was first discovered in 1820. And what tourmaline!

At twilight, which arrives early on late autumn days in the Oxford hills, two schoolboys, Elijah L. Hamlin and Ezekiel Holmes, were sauntering down the western slope of Paris Hill enjoying the glow of the setting sun on the flame of fall colors over the lofty, distant mountains.

Spellbound, they lingered to watch the purple shadows deepen in the valleys. As they turned to descend the hillock, young Hamlin was jolted from his trance by a vivid flash of green. Lying loose in the earth, clinging to the roots of a blown down tree, was a fragment of a transparent green crystal. They pocketed the beautiful piece and knelt to look for more. But, by now it had become dark and a hot supper beckoned, so they restrained their excitement and planned an early search before school the next morning.

During the night, a heavy Maine snowstorm blanketed the hills. The deep drifts through which the boys trudged to school the next morning remained until spring! Imagine their impatience throughout that long winter!

The first thaw sent them racing to the outcrop on the summit of the hill. What they found made up for every suspenseful minute of their wait. Lying on top of the bare ledge, sparkling in the sun, were dozens of crystals in all the gay colors of the most bewitching of all gems — the tourmaline.

You can see some of these large, limpid Mt. Mica crystals in museums near and far: In Maine, in Massachusetts at Harvard's Peabody Museum, in New York at The Museum of Natural History, in Washington, D. C. at The Smithsonian, in Chicago at The Field Museum, in Austria at The Hapsburg Museum, in Germany at The Kaiser's Museum, in Sweden's Royal Museum; and, if you can get in and can locate the late Czar's crown, right in the center of the diadem. This famed central stone was thought by the Russians to be an enormous, flawless ruby and by the old timers in Oxford County to be their lost or stolen rubellite (tourmaline like unto a ruby) which mysteriously disappeared on its way to Washington.

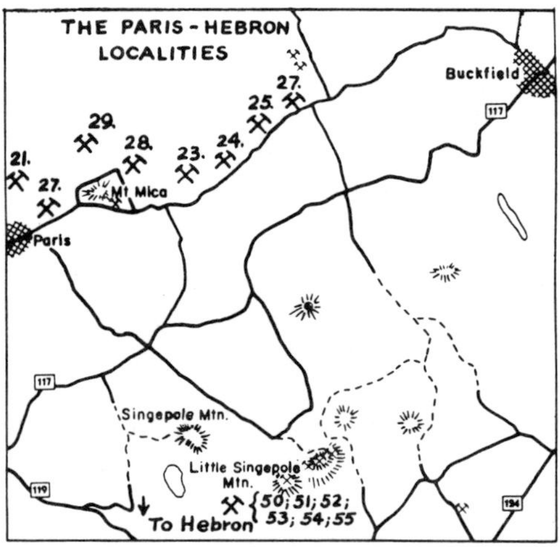

Map 28—Mt. Mica 2 miles from Paris Village on road to N. Buckfield. Take l. turn dirt road ½ mile.

The story they still tell is that when the news of the fabulous find reached Washington, D. C., perhaps through Oxford County, Maine's Hannibal Hamlin, Vice President under Lincoln, statesmen, foreign ambassadors and gem experts flocked to Paris Hill.

Baron Loederer, the Austrian ambassador, and Mr. Cramer, the Russian, were frequently seen searching there, intensely interested.

The best crystals were intrusted to Abraham Lincoln to take to Washington for exhibition.

Somewhere between Maine and New Haven, Connecticut, jogging ahorseback over the rutted, deeply muddy roads, he lost one of the precious packages. Among other beauties, this package contained the rubellite which later showed up in the Czar's crown and an immense, double terminated green crystal which can be seen in Vienna today labelled "from the USA."

Map 29 . . . *Ryerson Hill Mine.* Owned by Jacob Immonen, West Paris. Reported: gem beryl, columbite and feldspar.

Scott Colby Mine. Owned by Eino Pulkinnen, South Paris. Reported: feldspar and rose quartz.

Slattery Mine. Owned by Roscoe Slattery, Paris, or the late Stanley I. Perham, West Paris. Reported: bertrandite, beryl, biotite, feldspar and rose quartz.

Stearns Farm Mine. Owned by Austin Stearns, Paris Hill. Reported: rose quartz (fusable for optical purposes).

Stony Brook Mine. Owned by the estate of Howard Irish. Reported: feldspar, lepidolite, muscovite, quartz, and tourmaline (damourite).

The first discovery of molybdenum, called "white gold" because it is so rare and valuable, was made above Paris Hill. Attention was called to the area when workmen who were laying water mains were found to be greasing their boots with a black substance found to be graphite.

PERU TOWNSHIP

Map 30 . . . *Hedgehog Hill Mine.* Owned by the estate of Howard Irish. Reported: gem beryl (pink morganite) and golden (heliodore), biotite, muscovite, chrysoberyl and garnet.

Tumbledown Dickison Old Mica Mine. Reported: beryl (golden and emerald), garnet, smoky quartz, feldspar, mica, purpurite and sylvanite.

Lobikis Mine. Owned by Mrs. John Lobikis, Rumford. Reported: beryl, muscovite, pyrrhotite, quartz, triphylite and vivianite.

Dixfield Quadrangle

PERU TOWNSHIP

Map 31 . . . *Perry Prospect.* Owned by Arthur Lemay, West Peru. Reported: autunite, petalite, spodumene and triphylite.

Fryeburg Quadrangle

Map 32 . . . At Center Lovell on the west side of the east slope of Sabattus Mountain is "The Devil's Staircase." This is a perpendicular cliff of micaceous rock in which mount two hundred and fifty natural rock steps.

DENMARK TOWNSHIP

Map 33 . . . *Warren Quarry (Howe Mine).* Owned by The Warren Lumber Co., Denmark. Reported: amethyst.

The late George R. Howe of Norway, first owner of this mine, was one of the best loved of the many loveable personalities connected with Maine minerals. Several mineralogists in the State are spoken of as George Howe's boys. He was generous to the youth of Oxford County with his knowledge and his time.

It was George Howe who found royal purple amethysts of finest water on Pleasant Mountain near Denmark. He found them in a vein of milky white quartz running through coarse granite. The choicest he found loose in a pocket imbedded in kaolin.

Let George Howe speak for himself, "Listen, I have jewels hidden in this place that even your mother has never seen! I have sapphires big as eggs and as blue as cornflowers. I have chrysolites and beryls and chrysoprases and rubies — no, this is not an emerald, though many, unfamiliar with the Maine tourmaline, might mistake it for one. Maine tourmalines are famous, you know, and are to be found in most of the large gem collections in the United States and abroad. The largest tourmaline ever found at Mt. Mica on Paris Hill is valued at well into six figures. It weighs 411 carats uncut, is flawless and the equal of the finest emerald in hardness and color."

He then held up a deep purple stone which glowed a rich burgundy red.

"This is a royal purple amethyst from Mt. Pleasant in Denmark. The most perfect found there was considered the finest royal purple amethyst in existence. At one time it reposed in the collection of the former Kaiser Wilhelm who valued it at $10,000.00. Lesser amethysts from Maine are in museums all over the world."

He went on to show to his entranced listeners his collection of gloriously beautiful Maine gemstones, aquamarines, golden beryls, cesium beryls sparkling like diamonds, topazes, smoky quartzes, moonstones, and fresh water pearls, some equal in value to the oriental, which he had extracted from mollusks in Maine trout streams.

Taking up a book on pearls and other gems written by the noted gem expert, Dr. George Frederick Kunz of Tiffany's, he opened it to a full page illustration of Maine's own, the singer Nordica, wearing the pearl necklace which always completed her costumes in her appearances all over the world.

Mr. Howe felt sure some, or all, of the pearls had come from Maine.

Only certain large, dark-shelled clams in certain glacial streams contain pearls. Most of the salt water mussels do, sometimes as many as twenty in one shellfish, usually very small.

LOVELL TOWNSHIP

Map 34 . . . *Forks Farm.* Owned by Henry Forks, Lovell. Reported: beryl, feldspar and muscovite.

Map 35 . . . *Maxim Prospect.* Owned by Alton Maxim, South Paris. Reported: feldspar, muscovite and quartz.

Maxim — there is an old Maine name well known from the Maxim who invented the gun silencer. When I asked the late Howard Irish, judging by his round ruddy face, light bushy hair, bright blue eyes and what seemed to me like a foreign accent, if he was a Swede he replied, "Nope, straight from the Mayflower through the Maxims on my mother's side. We've lived here on Paris Hill on one side or the other of this road, for three hundred years."

STONEHAM TOWNSHIP

From here through Sugarloaf Mountain, Maine's large ski resort, to Stowe is soda-feldspar ledge, different from most of the other granites in Maine which tend to be potash-feldspar rich.

It is in this type of pegmatite that we are likely to find topaz and amethyst. A characteristic by which you may recognize the soda feldspar, plagioclase, is the presence of striations on the most prominent cleavage faces. This is due to "twinning." At East Stoneham, 4.7 miles from the road, Rupert Aldrich found one of the best crystals of aquamarine ever known. See, the old days of the fabulous gem finds aren't gone forever!

Map 36 . . . *Foster Hill Prospect.* Owned by James Merrill, East Stoneham. Reported: feldspar and muscovite.

Lord's Hill Mine. See Map 9 . . . Listed under Bethel Quadrangle, Stoneham Township.

STOW TOWNSHIP

Map 37 . . . *Colton Hill Prospect.* Owned by John Powers, Univ. of Maine. Reported: large pegmatite (feldspar and quartz).

Map 38 . . . *Big Deer Hill Area.* Owned by Harry Eastman, Western Maine Forest Nursery, Fryeburg. Reported: quartz (amethyst, crystal and smoky), feldspar, garnet, muscovite and pyrite.

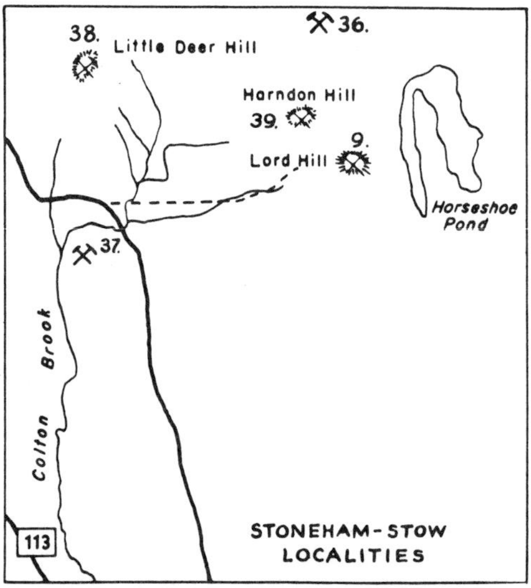

In and near White Mtn. National Forest. Route 113 N. 15 miles from Fryeburg to Stow. E. at Stow, across Cold River Bridge 3 miles to end. L. 4 miles to a fork. L. fork for 5 miles to quarries—away from road.

Here, and on Little Deer Hill, is the most promising amethyst location in Maine. They are usually found in pockets in the cliff face. You can detect the presence of pockets by pale amethyst and milky white quartz in the ledge. Back of this, near the surface, are sometimes small pockets with mica and quartz showing. By following inside these pockets with your geologist's pick you will sometimes open up larger pockets with white, clear quartz and smoky quartz crystals growing out of the walls and amethyst crystals inside. Former State geologist, the late Freeman Burr, used to find amethysts here in rotten pegmatite and in the clay overburden of the hillside.

Map 39 . . . *Harndon Hill.* Reported: Apatite, beryl, cleavelandite, columbite (one pocket afforded forty pounds of pure material), purple fluorite fills the small cavities in the cleavelandite. Striking specimens for a collection are pink to deep red botryoidal masses of kaolin-montmorillonite resembling rhodochrosite formed on crystals of cleavelandite. This highly mineralized hill produces triphylite, sometimes in huge masses, herderite in short prisms, bertrandite in pockets, autunite, biotite, gehlenite, garnet, muscovite, quartz and zircon.

This is a well known topaz locale. The late Dr. Kunz of Tiffany's wrote, "Harndon Hill was the first in New England to furnish good, clear and distinct crystals of topaz. They are the best crystals to be found in the United States. The finest are colorless or faintly tinted with green or blue. Less perfect ones are more decided in color, light shades of green, yellow or blue."

There is a radioactive ledge on the slope of Sugarloaf Mountain, the western part of Stoneham and the northwest slope of The Nubble. It is also part of the topaz region. Charlie Bragg has sometimes found topaz inside a big ball of yellow mica or wedged in between sheets of mica.

Dr. Kunz tells of two fine aquamarines found by farmers several miles from Sugarloaf. They were the finest in the United States and would have been the finest in the world had not the two farmers carelessly thrown down the crystals and broken them. We are so casual here in Maine regarding our world-shaking assets! Dana Douglas please note! A beautiful 133 carat gem cut from one of these crystals.

Freeman Burr, Charles Marble and Phil Morrill, all Maine mineral big shots, had a silly little experience here. They were stumbling down the mountain with heavy bags of topaz on their backs. Charles Marble was leading, head down and bent nearly double under his pack. His small cocker spaniel began to bark wildly.

Charlie Marble yelled at him to keep still but didn't bother to raise his head until he heard Phil Morrill burst out laughing. He had watched a full-grown black bear cross right under Charlie's unsuspecting nose with the tiny cocker in mad pursuit.

WATERFORD TOWNSHIP

Map 40 . . . *Beech Hill Mine (Waterford Mica Mine).* Owned by D. R. Kimball and G. Packard, South Waterford. Reported: biotite, muscovite, feldspar (microcline), garnet and quartz.

Knight Mine (Coye or Foye Mine). Owned by Winn Knight, Norway. Reported: beryl, columbite and muscovite.

Saunders Mine. Owned by C. P. Saunders, Waterford. Reported: chalcopyrite, feldspar, garnet, muscovite, magnetite, pyrrhotite, quartz and black tourmaline.

Map 41 . . . *Stearns Hill Mine.* Owned by Burton Newton, Bethel. Reported: feldspar, tourmaline, aquamarine, garnet, dendrites in feldspar, green apatite and pyrites. A 6x23-foot beryl crystal was found here. When the foreman put in a heavy charge of dynamite to get it out, it broke in three pieces. So, he just threw it on the dump, where presumably it still is today. Ah, me! Once again it happened!

Kezar Falls Quadrangle

DENMARK TOWNSHIP

Map 42 . . . *Lord Prospect.* Owned by Kenneth Lord, Denmark. Reported: biotite, muscovite, feldspar and quartz.

HIRAM TOWNSHIP

Map 43 . . . *Cutler Mountain (Frenchman's Gold Mine).* Owned by Lester Hammond, Hiram. Reported: feldspar, muscovite and quartz.

Norway Quadrangle

NORWAY TOWNSHIP

Map 44 . . . *Tubbs Ledge.* Owned by Andrew Eastman, Norway. Reported: cassiterite, lepidolite, rose quartz, green tourmaline and triphylite.

Cobble Hill. Reported: ledge with small but perfect crystals of chrysoberyl, zinc, spinel and zircon.

WATERFORD TOWNSHIP

Map 45 . . . *South Waterford Prospect (Bear Mt. Prospect).* Owned by Dr. R. E. Hubbard, South Waterford. Reported: muscovite.

Old Speck Mountain Quadrangle

Up this mountain to get to the gorgeous falls, Screw Auger and on above, Mother Walker, where there is the highest glacial cirque in the east, was where we followed the blue paint trail which was painted to lead us away from the treasure. The treasure we were seeking that day was the fine smoky citrine to be found under the tumbling water and nearby the falls and the pocket of large quartz crystals the Stearns Bryants found upstream from the right hand side of the bridge over Moss Brook.

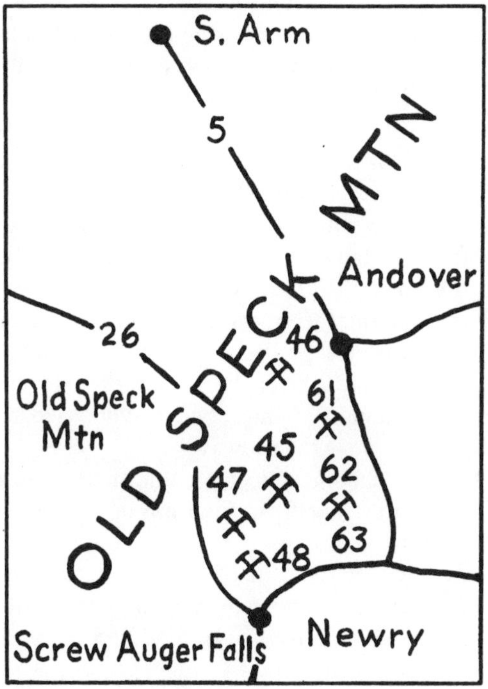

ANDOVER NORTH TOWNSHIP

Map 45 . . . *Spruce Mountain Area.* Reported: massive beryl.

ANDOVER WEST — SURPLUS TOWNSHIP

Map 46 . . . *Baldpate Mountain Area.* Reported: aquamarine beryl.

ANDOVER WEST — WEST SURPLUS

Map 47 . . . *Frye Brook-West Branch.* Reported: beryl.

NEWRY TOWNSHIP

Map 48 . . . *Puzzle Mountain.* Reported: beryl.

Oquossoc Quadrangle

Map 49 . . . *Umbagog Lake.* Reported: asbestos under the entire lake.

Helen Caldwell Cushman tells a stirring tale, **The Indians Were Here,** about the Lone Indian of the Magalloway, Metalluk, who lived most of his life on Lake Umbagog, one of the few lakes which still retains its original Indian name.

The Indians never forget a favor or a wrong.

The brave, Metalluk of the Cooashaukee tribe, meaning Dweller in the Pine Tree Country, should have been a chief of his tribe but was disqualified for rescuing a white girl from torture.

When his tribe was driven out of their hunting and fishing grounds by the English, he vowed to the Great Spirit that he would accept all hardships and remain on the hunting grounds of his forefathers.

Metalluk was a famous guide, skilled hunter and trapper and a fashioner of superbly made birch bark canoes. He was a friend of Governor Lincoln and many Congressmen who used to come to see him at his camp on the Magalloway.

When his beloved second wife Oozalluc died, he wrapped her in birch bark and suspended her from the limb of a tree that leaned over the smoke hole of his camp and thus embalmed her.

Ever since he died at the age of one hundred and twenty years there are those few, and only those in canoes, who catch an occasional glimpse of a ghostly white birch canoe paddled by a tall figure very rapidly, very expertly, very silently.

The canoe has certain distinguishing features which mark it as one built by Metalluk. Whenever it is seen there is always a loon nearby uttering its cry of desolate loneliness.

Those rare ones who have seen it become very lucky fishermen, no matter how poor their catch before. Those who land prize trout or salmon claim to have been helped by Metalluk.

The line at the beginning of Helen's story is from this poem by the late Stephen Vincent, given in part here:

I don't know why this Indian is,
A bow within his hand,
But he is hiding by a tree
And watching white men land.
They'll kill his deer and net his fish
And clear away his wood
And frequently remark to him
They do it for his good.
Then he will scalp and he will shoot
And he will burn and slay
And break the treaties he has made —
And, children, so will they.
But just remember this about
Our ancestors so dear:
They didn't find an empty land.
THE INDIANS WERE HERE.

Poland Quadrangle

HEBRON TOWNSHIP

Here is where the late Perien Dudley made the first find of pollucite, the chief cesium-bearing mineral, in this country.

Map 50 . . . *Barrett Prospect.* Owned by J. L. Barrett, Hebron. Reported: beryl, feldspar, garnet, muscovite and black tourmaline.

Conant Mine. Owned by Bell Minerals Co., West Paris.

Map 51 . . . *Hibbs Quarries.* Owned by Maine Mica Co., Spruce Pine, N. C., 1.3 miles west of Hebron village. Reported: beryl, biotite, muscovite, feldspar, garnet, quartz, black tourmaline, cassiterite and the deepest pink rose quartz in Maine.

Hill No. 4 Quarry. Owned by John Cummings, Paris. Reported: cassiterite, feldspar and gem tourmaline.

Map 52 . . . *Rubellite Mine (Cushman Mine).* Owned by Dwight Cushman, Hebron. Leased to Priscilla Bryant, Winthrop. Reported: gem tourmaline (rubellite and green), pollucite and beryl.

Map 53 . . . *Streaked Mountain Prospect.* Reported: feldspar.

Sturtevant Mine (Sanitarium Mine). Owned by Norman Sturtevant, West Minot or Bell Minerals Co., West Paris. Reported: beryl.

Map 54 . . . *Upper Rubellite Mine.* Owned by Dwight Cushman, Hebron. Leased to Priscilla Bryant, Winthrop. Reported: gem tourmaline (rubellite and green), smoky quartz crystals, beryl and pollucite.

This mine has caused a stir.

One of the first finds was a pocket of nice smoky quartz crystals. When Priscilla Bryant and her late husband Stearns brought them home they just happened to put a few under their ultra-violet light. And, lo and behold, they were fluorescent! Surprise, because this mine had not been known to have fluorescent minerals.

Just for fun, Stearns, who was very ingenious, rigged up a home-made portable black light and back they went to the mine at night. They were rocked back on their heels at the sight. The whole mine fluoresced!

Of the variety of material they brought home, each of which glowed a different vibrant color, yellow, green, blue or red, under the black light there were three they could not identify.

Off they sent them to the geology laboratory at Dartmouth College. No dice! No one could identify them. Then Harvard University wrestled with the three mystery minerals. It would be exciting to think that Maine had given three new minerals to the world, but the experts have decided to consider them just unusual varieties of feldspar.

DIRECTIONS:

To quarries in hills N.W. of Hebron take road S.E. of Paris Hill (near Paris Village). Cross Stony Brook 7 miles out of Paris. At sharp l. turn at a high hill on the r., take dirt road leading off to r., by a farm and up a hill to the quarry area. Well marked trail leads to the mine.

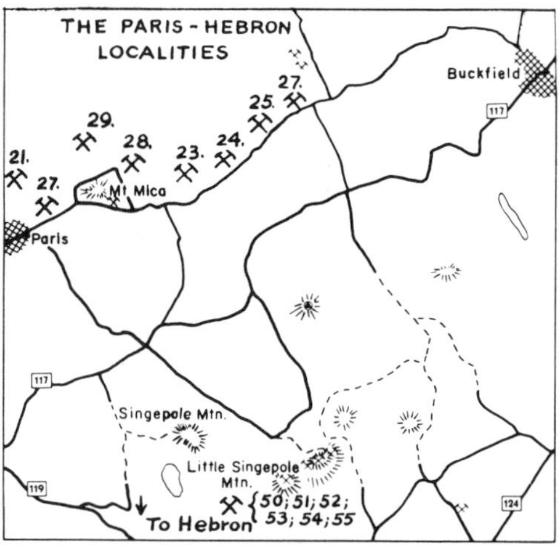

Map 28—Mt. Mica 2 miles from Paris Village on road to N. Buckfield. Take l. turn dirt road ½ mile.

Map 23—Bennett 4 miles beyond Mt. Mica Mine.

Map 54—Upper Rubellite 2 miles N. of Central Maine Sanatorium on Route 119-A. On steep ridge on r. N. of church.

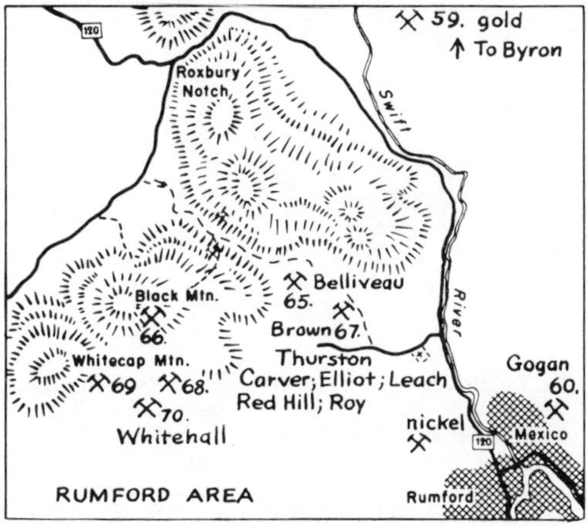

DIRECTIONS:

N. of Rumford on Rt. 120, along Swift River for 10 miles. At Roxbury Notch l. on dirt road into hills. 2 miles into mine road leading off l. towards mines. Mine road is car passable almost to largest quarry.

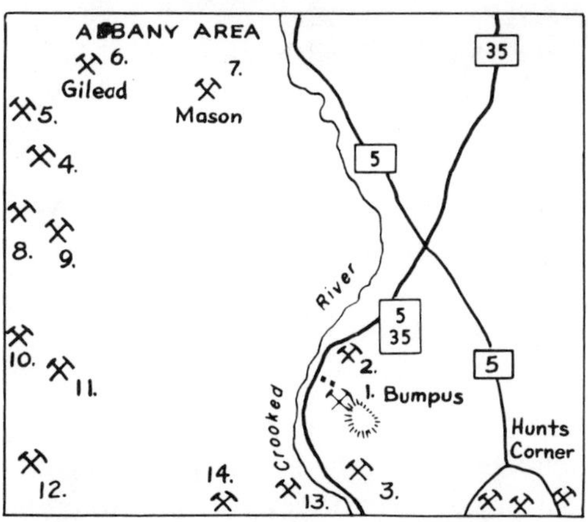

Map 1—Bumpus. 15 miles S. of Bethel on Route 5.

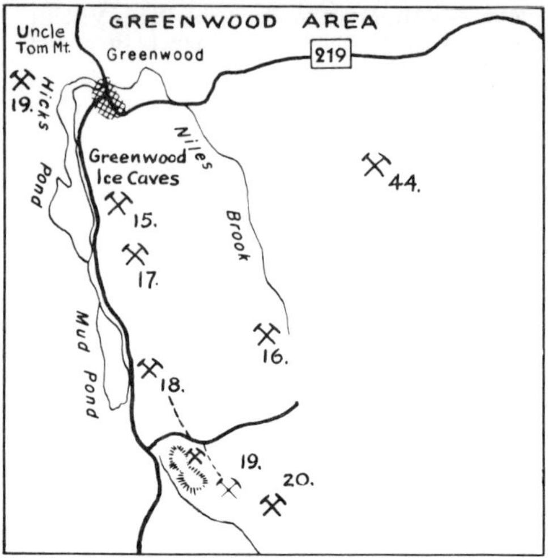

W. on Routes 117-118 out of Norway Village. Around S. end Pennesseewassee Lake. Take v. fork opposite 118 about 1/3 way up lake at intersection triangle. 6 miles to unimproved section. R. road climbing steep hill. Map 16; BB #7; 17—Diamond Ledge; 18—Harund (Noyes Mt.) and 19—Tanumnen and others are in this vicinity.

NOBLES CORNER TOWNSHIP

Map 55 . . . *Noble's Corner Mine.* Owned by the T. C. Mining Co. Reported: 16,000 grams of cesium, one of the largest finds ever.

OXFORD TOWNSHIP

Map 56 . . . *Jordan Prospect.* Owned by William Jordan, Mechanic Falls. Reported: feldspar, muscovite and quartz.

Rangeley Quadrangle

HOUGHTON TOWNSHIP

Map 57 . . . *False Gold Mine.* Owners jailed. Long since gone to the fiery region of forked tails and cloven hoofs. Reported: gold, but only what you can pan out of the Swift River. Here is where a mother and her two daughters have been getting their pocket money for years out of the gold they pan and sluice.

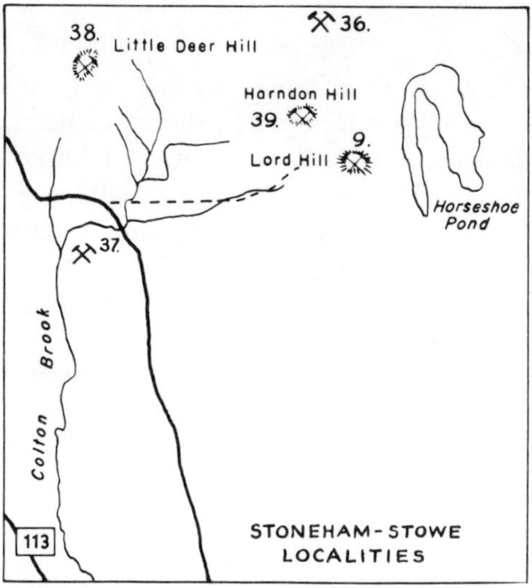

In and near White Mtn. National Forest. Route 113 N. 15 miles from Fryeburg to Stow. E. at Stow, across Cold River Bridge 3 miles to end. L. 4 miles to a fork. L. fork for 5 miles to quarries—away from road.

MADRID TOWNSHIP

Map 58 . . . *Niles Brook.* Reported: gold. Here is where the Jake the Trapper used to pan the nuggets which were found to be half gold and half platinum.

Not all of the gold hereabouts is alluvial. Rock mining for gold is carried out in the Rangeley area in a volcanic plug containing a dense black rock.

Rumford Quadrangle

BYRON TOWNSHIP

Map 59 . . . *The Swift River.* Reported: gold. I reported this fully in Chapter 3. Also found here: scheelite (tungsten ore), garnet, diopside and pyrite.

MEXICO TOWNSHIP

Map 60 . . . *Gogan Prospect.* Owned by Maurice Gogan, Rumford. Reported: feldspar and muscovite.

Map 61 . . . *Dunton Mine (Newry Gem Mine).* Owned by Plumbago Mining Co. Reported: amblygonite, apatite, autunite, beryl, beryllonite, cassiterite, chalcedony, cookeite, eosphorite, feldspar, francolite, herderits, lepidolite, microlite, montebrasite, muscovite, opal, pollucite, pyrite, quartz, redingite, siderite, sphalerite, spodumene, torbernite, gem tourmaline of all colors, triphylite, uraninite, vivianite and zircon. You name it. Newry has it. This is the famous locale our Maine geologists say is full of oddballs and far-outs.

Map 62 . . . *Newry Mines (False, Lower Mine, Nevel Quarry:* Dick Nevel, who owned a mineral shop in South Paris and owned this mine was killed here by dynamite). General Electric and Westinghouse have mined here for pollucite and cesium. Canadian owned. Reported: much the same group of far-outs at the Dunton Mine: amblygonite, apatite, autunite, beryl, beryllonite, cassiterite, columbite, eosphorite, fairfieldite, feldspar (cleavelandite), galena, garnet, hatchettolite, herderite, lepidolite, muscovite, pollucite, purpurite, quartz, siderite, sphalerite, spodumene, tantalite, torbernite, gem tourmaline of all the colors, triphylite, uraninite and vivianite. This list has a slight edge over the other mine and at that neither list is complete. How about yttrocerite, uranopilite, uraninite, stewartite, pitchblende, microlite, manganite, alunogen, childrenite, cimolite, columbite, and ilmenite and probably several more I haven't thought of!

The late Stan Perham told me what should encourage those of us who go out time and again and come back with very little to show for our trip. Every once in awhile someone hits the jackpot. And one of these times it could be you. A group of scouts from New York stopped in his mineral store in West Paris. One of them asked, "Is it true that a real emerald was found at Newry years ago?" "Yes," answered Stan. "Well, Mr. Perham, could you please tell me exactly where to look so I can find one, too?"

Stan, highly amused, told him to look along the stone wall on the northwest side of the upper dumps. Then, to look closely for sheets of mica between the layers of which he just might find an emerald.

The kid came back late that afternoon and said, deadly serious, "I went exactly where you told me and look —."

He held out a sheet of mica with, between the layers, a large, fine crystal of emerald! Stan said if he'd known that, he would have gone there himself.

Map 63 . . . *Plumbago Mountain and Scotty Mine.* Owned by International Paper Co., Livermore Falls, or Whitehall Co., 17 Battery Place, N. Y. Reported: beryl, beryllium and spodumene. Nice micromounts and rose quartz crystals. What, no plumbago (the old name for graphite)!

The last time I was there, The Beryllium Corp. of America was blasting the top off the mountain.

The first time I was there I climbed it by mistake trying to find Newry. A mile up the wrong mountain, all alone, too. A mile down again, where I shuddered to see fresh bear tracks on the path to my jeep. Then, another mile up the right mountain to the Newry Mines and back down again in the dark lugging a hundred pounds of rock. And I still love rockhounding!

ROXBURY TOWNSHIP

Placer mining for gold here.

Map 64 . . . *Binford Prospect.* Owned by Victor Binford, Roxbury. Reported: muscovite.

Partridge Peak. Reported: scheelite (tungsten). Small pinpoint grains in lime-silicate quartzite on west side of peak.

RUMFORD TOWNSHIP

Map 65 . . . *Belliveau Prospect.* Owned by Alfred J. Boucher, Isthmus Road, Rumford. Reported: biotite, muscovite, feldspar (microcline), garnet and quartz.

Map 66 . . . *Black Mountain Quarries.* Reported: apatite, beryl, cassiterite, columbite, garnet, feldspar (cleavelandite), hornblende, lepidolite, magnetite, muscovite, purpurite, quartz, sphene, spodumene, tantalite, tourmaline and uraninite.

This is another major far-out locale for exotic beatnik minerals. With three uranium minerals listed, hornblende, tantalite and uraninite, it is no wonder my geiger counter gave me a thrill by buzzing eight above the ground count!

Above the mines on the top of the mountain occurs sugar-white beryl, and nowhere else on the face of the earth!

This book would have been twice as thick if I had mentioned the scenic beauty of most of the places in this favored State of Maine I have recommended for your treasure hunting.

But, a few words about Black Mountain. It is on the far eastern edge of the White Mountain range so expect to be overcome by the sunset effects of The Presidential Range to the west. In the mines, themselves, the fan formations of the pink tourmaline crystals are pretty as is the lilac lepidolite spreading over the entire base of the hillside mine dumps. It looks like a field of wild asters.

Map 67 . . . *Brown-Thurston Prospect.* Owned by Mrs. Tom Brown, Bethel, or Lee Thurston, Rumford Center. Reported: beryl, cassiterite, feldspar (cleavelandite), columbite, garnet, muscovite, quartz and tourmaline.

Map 68 . . . *Carver Prospect.* Owned by Dana Putnam, Rumford Corner. Reported: beryl, feldspar and purpurite.

Elliott Mine. Owned by George Elliott, Rumford. Reported: apatite, beryl, feldspar, garnet, muscovite, quartz and black tourmaline.

Leach Quartz Prospect. Owned by George Crooker, South Andover. Reported: quartz crystals.

North Rumford Nickel Mine. Owned by Dean McCrillis, Byron.

After Mr. McCrillis retired, he returned to Maine and bought for $700.00 the nickel mine his grandfather had once owned. His interest grew until he now has a mineral shop in Byron and is cutting stones. His father, William, had been active in The Mendearth Co., producing "Nature's Plant Food" from the element-rich sandstone of the area. It is not producing now, but the shute all the way down the mountain can still be seen.

Map 68 . . . *Red Hill Mine.* Owned by The Whitehall Co., 17 Battery Place, N. Y. Reported: beryl, feldspar and triphylite.

Roy Mines Goddard Ledge). Owned by G. R. Alouisy, Coral Gables, Weymouth, Mass., or R. E. Roy, North Rumford. Reported: beryl, corundum, feldspar, pyrrhotite and sphalerite.

Map 69 . . . *Whitecap Mountain Area.* Owned by Thomas Dickson, Mexico. Reported: apatite, beryl, feldspar, garnet, muscovite, quartz and tourmaline.

Map 70 . . . *Whitehall Prospect.* Owned by Whitehall Co., 17 Battery Place, N. Y. Reported: beryl.

PENOBSCOT COUNTY

Bangor Quadrangle

BANGOR TOWNSHIP

Map 1 . . . *Queen City.* Reported: lead, silver and gold.

PENOBSCOT COUNTY

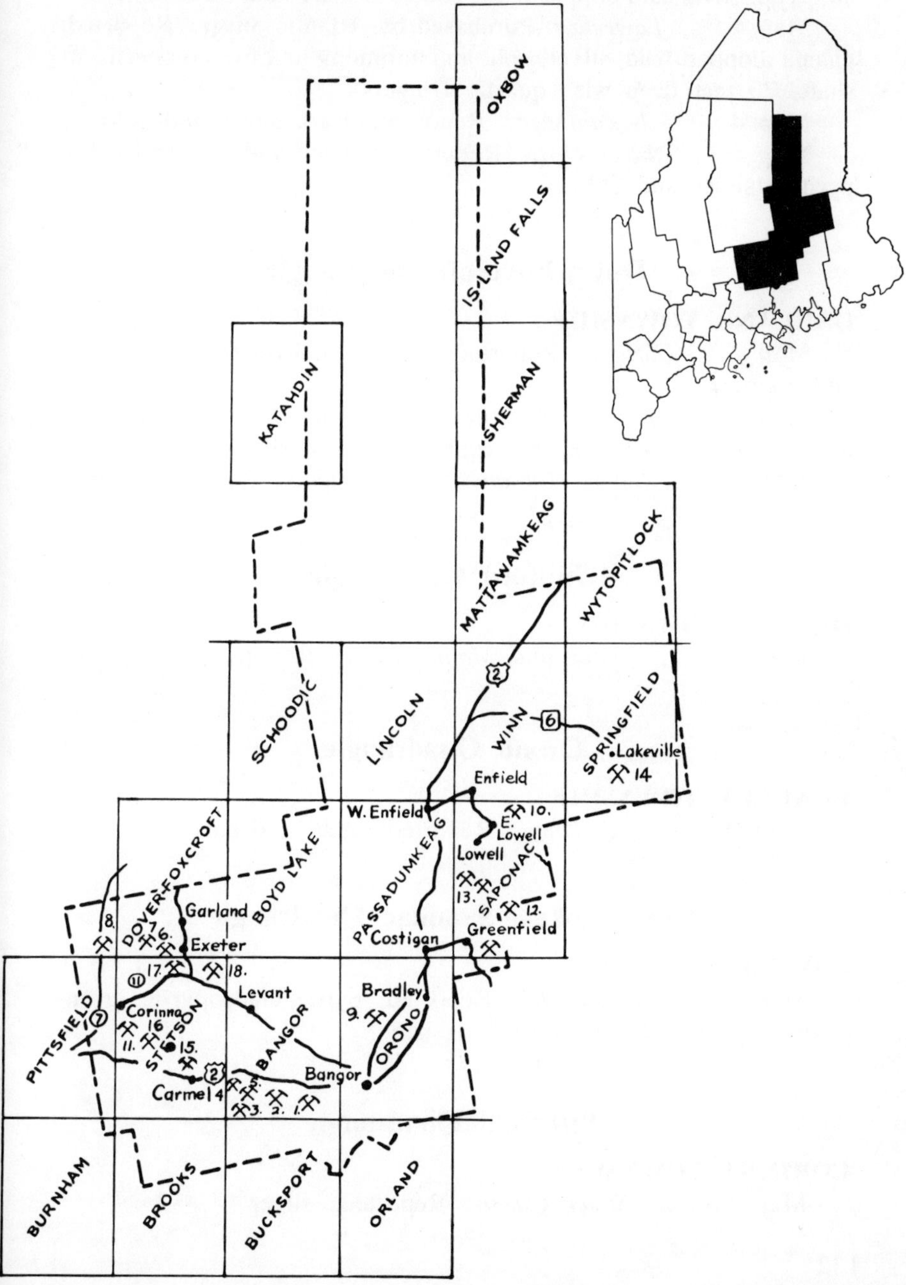

Map 2 . . . *Hampden (Consolidated Hampden, Dunton).* Reported: lead, silver and copper from 200-foot shaft sunk on a vein.

Map 4 . . . *Lawrence.* Purchased by Atlantic Silver. Reported: galena (copper, lead, silver, gold and antimony in a two compartment shaft, 70 feet deep with quartz stringers).

Map 4 . . . *Norumbega.* Reported: lead, silver and gold.

Map 5 . . . *Silver Drift.* Reported: galena and auriferous iron pyrites (silver and gold).

Dover-Foxcroft Quadrangle

GARLAND TOWNSHIP

Map 6 . . . *Garland.* Reported: galena, chalcopyrite (lead, silver and copper).

Map 7 . . . *Jones Hill.* Reported: galena (lead and silver) in quartz vein. Mass of galena running high in silver.

Map 7 . . . *Silver Hill and West Garland.* Reported: silver.

Guilford Quadrangle

DEXTER TOWNSHIP

Map 8 . . . *Dexter and Maine Silver.* Reported: silver.

Orono Quadrangle

BRADLEY TOWNSHIP

Map 9 . . . *Old Town.* Reported: gold and silver.

Passadumkeag-Saponac Quadrangle

LOWELL TOWNSHIP

Map 10 . . . *Vinegar Hill.* Reported: pyrite, chalcopyrite, arsenopyrite (gold and copper in two shafts).

Pittsfield Quadrangle

CORINNA TOWNSHIP

Map 11 . . . *Silver Queen.* Reported: silver.

Saponac Quadrangle

GREENFIELD TOWNSHIP

Map 12 . . . *Cemetery Hill and Hall Hill.* Reported: manganese and iron. Discovered in 1957 during field work by the Maine Geological Survey. Deposit strikes generally NE toward Passadumkeag Mountain, varies ⅛ to ¼ mile in width.

LOWELL TOWNSHIP

Map 13 . . . *Bassick and Shorey.* Reported: sulphurets (silver, antimony and gold) in a shaft 75 feet deep.

Springfield Quadrangle

LAKEVILLE TOWNSHIP

Map 11 . . . *Getchell Mt. Prospect.* Owned by Hubert Snow, Springfield, R.F.D. Reported: biotite, feldspar, magnetite and rose quartz.

Stetson Quadrangle

CARMEL TOWNSHIP

Map 15 . . . *Boyd, Carmel Antimony.* Reported: stibnite (antimony).

Harrington. Gold.

CORINNA TOWNSHIP

Map 16 . . . *Dearborn.* Reported: argentiferous galena (lead and silver).

EXETER TOWNSHIP

Map 17 . . . *Exeter.* Reported: lead, silver, galena and gold trace. 70-foot shaft.

LEVANT TOWNSHIP

Map 18 . . . *McKusick.* Reported: gold and silver.

PISCATAQUIS COUNTY

Dover-Foxcroft Quadrangle

Between Dover-Foxcroft and Milo on an outcropping of rocks north of the road are one of the few petroglyphs (ancient stone carvings) in Maine or New England. They are snowshoe-like webbed tracks evenly spaced four and a half feet apart for a distance of more than a hundred yards.

DOVER-FOXCROFT TOWNSHIP

Map 1 . . . *Foss Farm.* Reported: chalcopyrite (copper).
Map 2 . . . *Weston Farm.* Reported: chalcopyrite (copper).

Guilford Quadrangle

GUILFORD TOWNSHIP

Map 3 . . . *Guilford.* Reported: argentiferous galena, chalcopyrite (lead, copper and silver). There are fossils in the gravel pits beside the Piscataquis River.

Map 4 . . . *Monson Township.* Reported: slate. Used in New York's Empire State Building. The only slate in the nation which meets the high qualifications needed for electrical purposes.

Moosehead Lake Quadrangle

In 1952, a New York firm, Freeport Sulpher Co., leased large tracts of wild land from the paper company owners, Great Northern, Hollingsworth, and Whitney and Penobscot Development Company. The royalties start at $30,000.00, $10,000.00 for each of the three owning companies. They hope to recover iron sulphides.

Map 5 . . . *Old Squaw Mountain near Greenville Junction.* Reported: fossils.

This wide fossiliferous band extends from Somerset County near the Moose River, Jackman and Parlin Pond all the way across Piscataquis County to Millinocket, including the slopes of Old Squaw Mountain and Mt. Katahdin.

The north flank of Old Squaw Mountain and SE of Rockwood, a quarter of a mile south of the station where there is an abundance of fossils in certain layers of the railway cuts are hopeful hunting.

PISCATAQUIS COUNTY

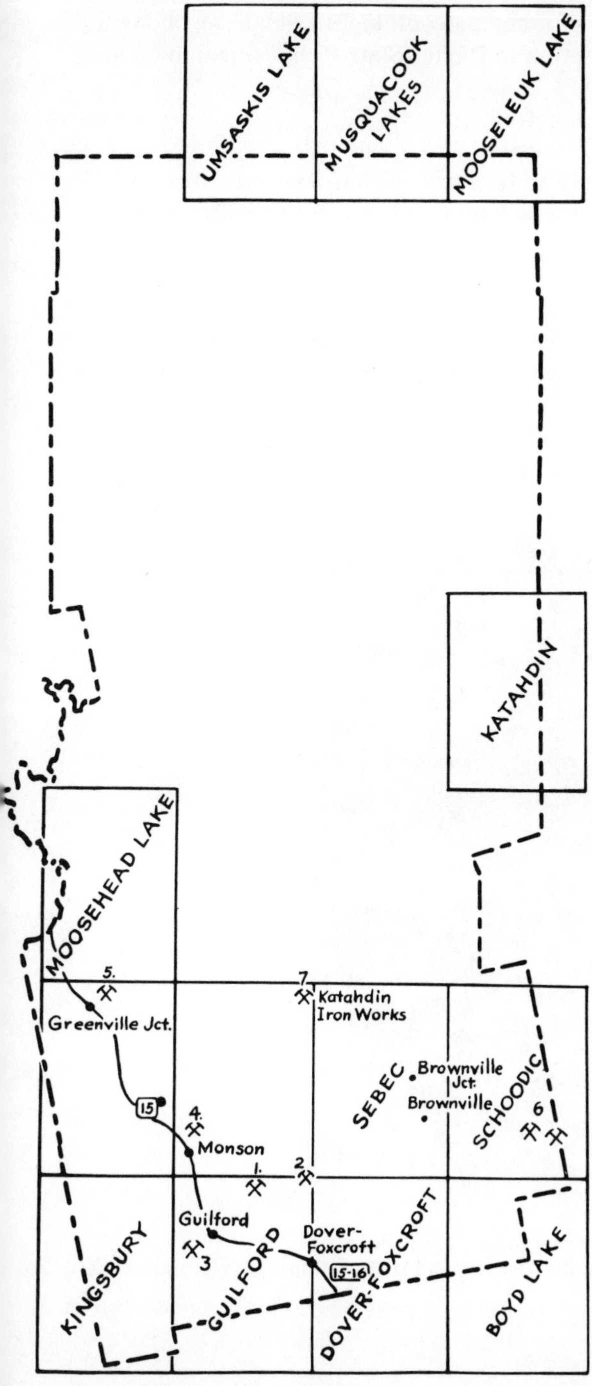

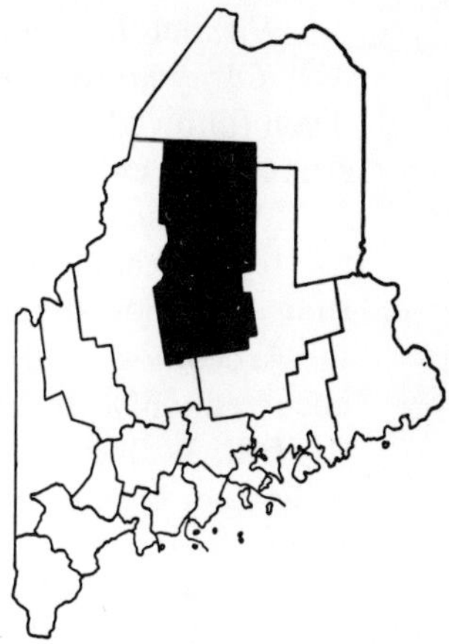

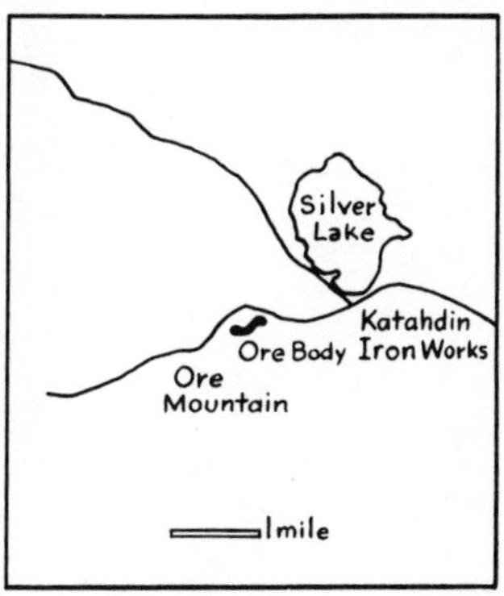

Not long ago a group of noted paleontologists wandering along the Trout River Valley Formation in Baxter State Park came upon a man digging.

"What are you looking for?"

Without looking up, "Fossils."

"Oh, you are not likely to find any in that lava rock formation." Then followed a learned discussion of where to find fossils, their classifications, etc.

"Well now — I've been having pretty good luck."

It took them down a peg or two when Bill showed them his fossil finds, superior to anything they had found. Some, which he had found in Aroostook and Somerset Counties, they had always thought occurred only in Europe.

Bill, William Forbes, Professor of Paleontology at the University of Maine, Orono, has had an original fossil find named after him.

Brachiopod fossils from the area of
Old Squaw and Katahdin Mountains

Sebec Quadrangle

BROWNVILLE TOWNSHIP

Map 6 . . . *Crocker Quarry on the east side and Merrill Quarry on the west side of the Pleasant River.* Reported: slate.

The mining of slate is one of Maine's oldest industries, begun in 1842. In 1876, Maine slate won awards at The Philadelphia Exposition.

The art of splitting slate has been handed down from generation to generation, some of the skilled artisans losing their lives from silicosis, inhaling slate dust.

General Electric has operated here for slate for insulating boards. Portland Slate Co. has the biggest and deepest tunnels.

The Brownville Slate Co., operating The Barnard Quarry, estimates the value of the slate contained in the five quarries here at $375,000,000.00. Today, Maine slate is used principally for making decorative pieces, such as lamps and furniture.

Map . . . *Katahdin Iron Works.* Owned by Allied Chemical Corp. Reported: pyrrhotite, chalcopyrite, ilmenite, magnetite, linnaeite, marcasite (sulphur and iron). Fossils.

Mt. Katahdin, highest in Maine, long thought by the Indians to be a man-god who sometimes woos maidens a la Greek gods on Olympus, is actually, one huge granite magma (a volcano that didn't quite make the grade).

Iron ore was first discovered here in 1843 by Moses Greenleaf. The deposit is roughly rectangular in shape, 200 feet long and an average of 400 feet wide. The host rock is massive, coarse-grained, dark gray to purplish gray plagioclase-rich norite. The sulphides in the mineralized zone increase from 3% to 80% near the center.

In 1952, General Chemical bought the old iron mine from the Piscataquis Iron Works Corp. to recover the strategic sulphur and iron. They have offered $1,200,000.00 for the shares of stock left in the will of F. W. Hill of Bangor, to The Bangor Public Library, Eastern Maine General Hospital, and The University of Maine. The State Park Commission has recently taken over the beehive iron smelters as an historical site.

SAGADAHOC COUNTY

Here is a rich area, one end of the pegmatite sheet which runs in a northwesterly direction through Androscoggin and Oxford Counties,

SAGADAHOC COUNTY

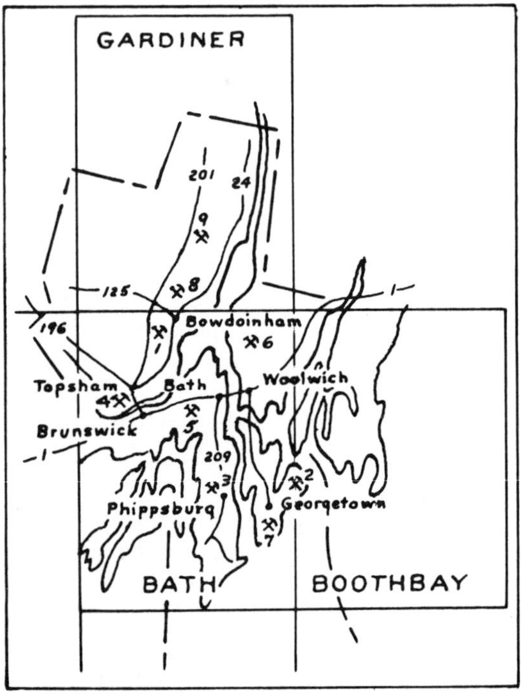

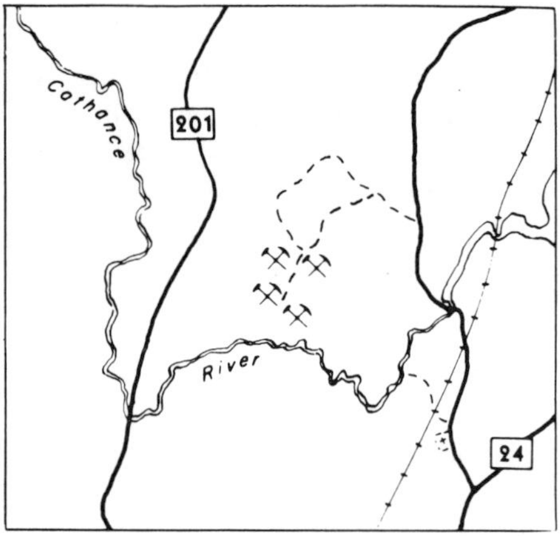

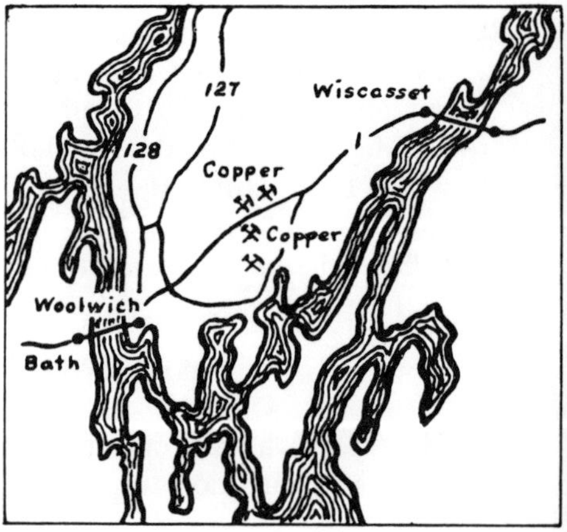

establishing its own world records on the way.

There has been·much talk of coal in this area, although some experts claim it is the wrong kind of rock for coal, so that any found must have washed ashore from ships.

Could be. I remember when a group of men from the Sample Shipyard here in Boothbay Harbor came to me in great excitement with a large lump of black rock to try my geiger counter on for radioactivity. When Luther Dunton saw it and heard they had found it on the shore of the Sheepscot River, he laughed fit to die. When he was a boy, his father, captain of a river steamer, used to race other steamers down river from Wiscasset. So the owners wouldn't know they had been wasting fuel getting up an extra head of steam, the captains used to dump the telltale slag on the shore just before putting into Boothbay Harbor. Luther recognized the large lump of, we hoped, "pitchblende," as one he had helped his father get rid of just before rounding Hendrick's Head Light.

But there are those who think that the coal found hereabouts could not have washed up.

In 1927, Mr. W. C. Davis, president of the Kennebec Refining Co., said there is enough coal within a short distance of Bath to supply all of New England; and that there is also paraffin oil. Lumps and, occasionally, large blocks wash ashore near the mouth of the Kennebec River and at Small Point.

A mining engineer is certain coal exists beneath Morse Mountain.

A Mrs. Stover found a HUGE lump of coal at Popham, too large to have washed ashore.

In 1928, the trawler *Alice M. Doughty* lifted her dragnet off Seguin Island to find it weighted down with hard anthracite coal. A week later the dragger *Barbara* brought home a bushel basket full of big pieces of hard coal dragged up from the same deep water, open ocean spot.

For many years, Seguin residents have discovered pieces of coal washed up after heavy storms.

Bath Quadrangle

BOWDOINHAM TOWNSHIP

Map 1 . . . *Adderton Quarry*. Owned by Mrs. Frank Adderton, Bowdoinham. Reported: feldspar.

Booker. Owned by Emory W. Booker, Brunswick. Reported: feldspar.

Consolidated Quarry. Owned by John Powers, Univ. of Maine, Orono. Reported: biotite, feldspar, garnet and magnetite.

Dunn Prospect. Owned by Kenneth Dunn, Bowdoinham. Reported: feldspar.

GEORGETOWN ISLAND TOWNSHIP

Map 2 . . . *Consolidated Quarry.* Owned by John Powers, Univ. of Maine, Orono. Reported: autunite, beryl, feldspar, gummite, lepidolite, muscovite, quartz and tourmaline.

Moffatt Mines. Owned by Elizabeth D. Moffatt, Georgetown. Reported: feldspar and tourmaline.

Thorne Prospect. Owned by Bruce Thorne, 93 Bay State Road, Melrose, Mass. Reported: biotite, feldspar, muscovite, quartz and black tourmaline.

Todd Mine. Owned by Will Todd, Georgetown. Reported: beryl and feldspar.

PHIPPSBURG TOWNSHIP

A mother lode of gold is said to be at Small Point, out on the tip.

Map 3 . . . *Basin.* Reported: feldspar, garnet and muscovite.

McKay Farm Prospect. Owned by Charles Beech or Alton Whittier, Brunswick. Reported: beryl, biotite, feldspar, garnet, muscovite, quartz and black tourmaline.

Parker Head. Reported: beryl, feldspar, ilmenite, magnetite and muscovite.

Perry Mine. Owned by Richard Hatch, Cundy's Harbor. Reported: feldspar.

Robbins Mine. Owned by Mel Robbins, Brunswick. Reported: beryl, columbite, feldspar and magnetite.

Rogers Mine. Owned by Dorothy Rogers, Winnegance. Reported: gem beryl, feldspar and black tourmaline.

Thomas Quarry. Owned by Atkins Bay Farm, Phippsburg. Reported: apatite, gem beryl, feldspar, muscovite and tourmaline.

TOPSHAM TOWNSHIP

This is the feldspar hub of Maine. Other things, too, though. The quartz for the 200-inch lens of Lick Observatory, the largest astronomical lens ever cast, is from Topsham.

Gemstones — Topsham produced the largest topaz ever discovered in North America and the largest and finest emeralds.

TOPSHAM AREA

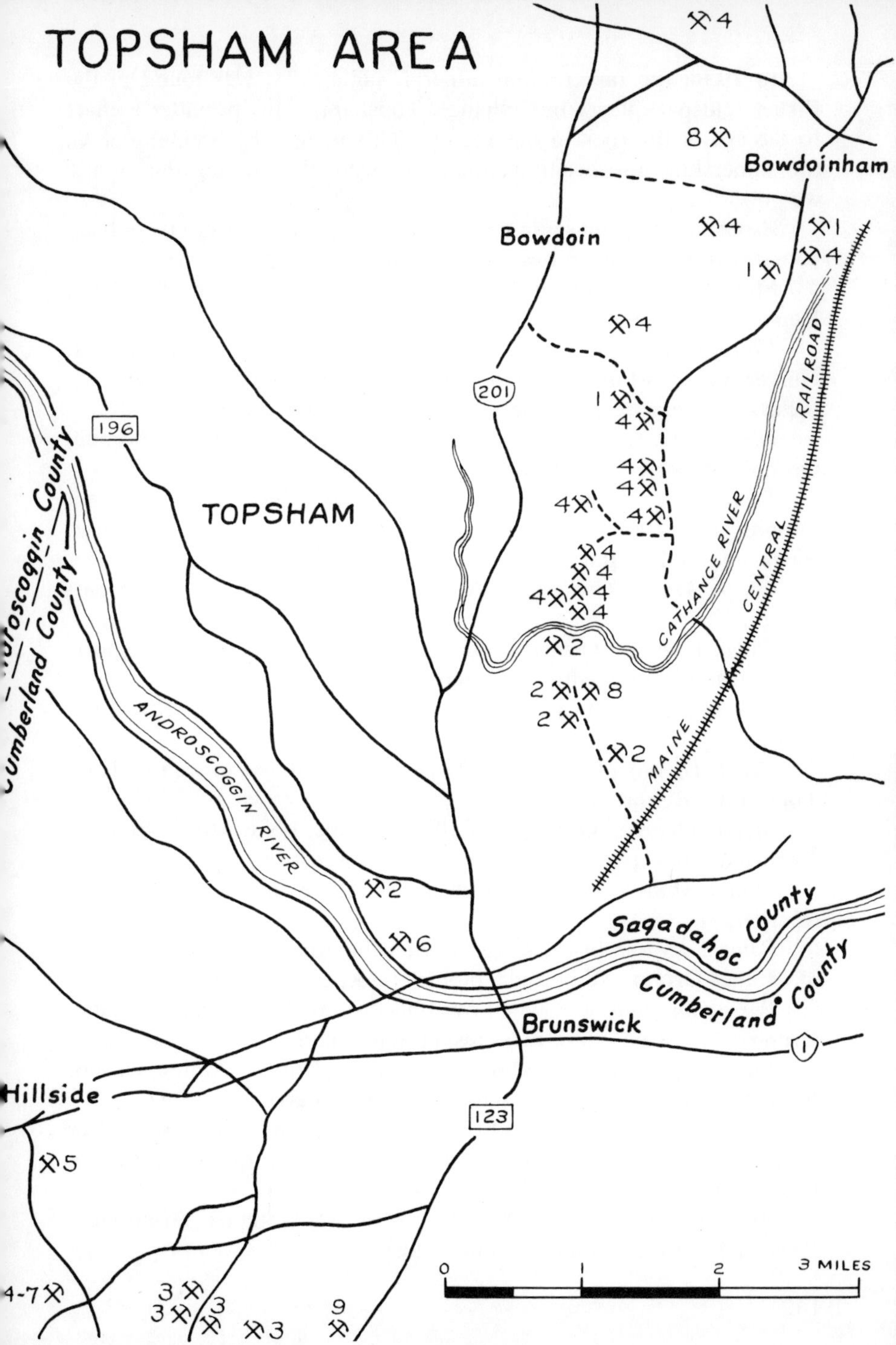

In 1934, the rare radium mineral, samerskite, was found in the Fisher feldspar quarry in Cathance, Topsham. This provides a chart to the age of the rock in this region. This is done by breaking down the samerskite into uranium, then into lead, then taking the atomic weights.

Especially fine museum specimen crystals of uraninite have been found in the various mines in and around the Fisher Quarry.

Map 4 . . . *Chapman.* Owned by Charles Chapman, Topsham. Reported: feldspar and quartz.

Consolidated Quarries Nox. 1, 2, 3 and 4. Owned by John Powers, University of Maine, Orono. Reported: feldspar, beryl, biotite, muscovite, quartz, gem tourmaline, garnet and magnetite.

Direnzo Prospect. Owned by John Powers, University of Maine, Orono. Reported: feldspar, muscovite and quartz.

Fisher Quarry. Owned by John Powers, University of Maine, Orono. Reported: feldspar, columbite, beryl, muscovite, quartz, topaz and tourmaline.

Given Mine. Owned by Arthur Graves, Topsham. Reported: feldspar.

Given Quarry. Owned by Cesare Trusiani, Brunswick. Reported: feldspar, muscovite and quartz.

Graves Mine. Owned by Arthur Graves, Topsham. Reported: feldspar.

Great Divide (Undivided Quarry). Owned by Maynard Tarr, Topsham. Reported: feldspar.

Ingalls Quarry. Owned by William Ingalls, Lake Alfred, Florida. Reported: feldspar (plagioclase) and muscovite.

Mallet Mine. Owned by Edward Elliott, Brunswick. Reported: feldspar.

Powers Mountain Prospect. Owned by John Powers, University of Maine, Orono. Reported: biotite, feldspar and quartz.

Purington Quarry. Owned by John Powers, University of Maine, Orono. Reported: biotite, feldspar and quartz.

Rumrill Quarry. Owned by John Powers, Univ. of Maine, Orono. Reported: beryl, feldspar, muscovite and quartz.

Russell Brothers Mine. Owned by A. R. Russell, Topsham. Reported: beryl, biotite, feldspar, garnet, muscovite, quartz and black tourmaline.

Alice Staples Quarry. Owned by Mrs. Alice Staples, Brunswick. Reported: beryl, columbite and feldspar.

William Staples Quarry. Owned by Russell Garland, Topsham. Reported: feldspar.

Trenton Mine. Owned by John Powers, University of Maine, Orono. Reported: feldspar, muscovite and quartz.

William Killes Quarry. Owned by Harriet Fisher, Brunswick. Reported: beryl, columbite and feldspar.

WEST BATH TOWNSHIP

Map 5 . . . *Davis Mine.* Owned by Mrs. Warren Davis, West Bath. Reported: apatite, beryl, feldspar (microcline), garnet, muscovite, pyrite, quartz and black tourmaline.

WOOLWICH TOWNSHIP

Amethysts have recently been found in a new road cut. New roads often open up promising hunting.

Map 6 . . . *Trott Cove Mine.* Owned by Sam Guptill, Woolwich. Reported: green apatite, beryl, feldspar, garnet, quartz, muscovite, black tourmaline, malachite, copper and recently some fine amethyst.

Boothbay Quadrangle

Map 7 . . . *Georgetown Island Township.* Formerly owned by Walter Cummings, Georgetown. Reported: feldspar.

This is the oldest mine in the State. The owner was telling me that when his father was operating it for feldspar, whose chief use is to glaze fine porcelain, he sent feldspar from his own mine to England to have a dinner service of finest china made for himself. Probably one of the big reasons feldspar mining has declined is — how many of us indulge in dinner services of fine china nowadays?

He told me his father was left-handed so ordered all the cups made with the handles on the opposite side. I swallowed that whole until my husband remarked, "What's to prevent a left-handed person from turning the cup around?"

Emery Booker of Brunswick who sold the old Smith mine to Earl Brown, told me the little island right off the stone dock used to yield fine tourmaline crystals. He said the Thomas mine he used to own is rich in carnetite and autunite, the uranium minerals.

Gardiner Quadrangle

BOWDOIN TOWNSHIP

Map 8 . . . *Coombs Quarry.* Owned by Benjamin Coombs, Cherry Street, Bath. Reported: beryl, biotite, feldspar, garnet, muscovite, pyrite, quartz and black tourmaline.

Ordway Mines. Owned by Mrs. Lilly Ordway, Bowdoin. Reported: beryl, biotite, feldspar, garnet, muscovite, pyrite, quartz and black tourmaline.

Trufant Prospect. Owned by U. S. Gypsum Co., Lisbon Falls. Reported: beryl, biotite, feldspar, garnet, muscovite and quartz.

BOWDOINHAM TOWNSHIP

Map 9 . . . *Consolidated.* Owned by John Powers, University of Maine, Orono. Reported: feldspar.

Harriman Mine. Owned by Harriman Construction Co., Bowdoinham. Reported: feldspar and quartz.

Lang Mine. Reported: feldspar.

McFee. Owned by Robert McFee, Bowdoin. Reported: feldspar.

SOMERSET COUNTY

Anson Quadrangle

EMDEN TOWNSHIP

Here is a petroglyph known locally as Indian Ledge. It is ¼ mile south of the Salem bridge on Route 201A, accessible by a footpath. The 50-foot hard shale ledge, which extends into the Kennebec River, is profusely covered with pecked out pictures, some suggesting contact with the early French trappers and missionaries.

Bingham Quadrangle

CONCORD TOWNSHIP

Map 1 . . . *Robinson.* Reported: pyrite, galena, sphalerite, chalcopyrite, arsenopyrite, bornite and pyrrhotite. The ore body is from 2 to 8 feet wide.

The Forks Quadrangle

THE FORKS — EAST MOXIE

Map 2 . . . *Moxie Pond (Black Narrows).* Reported: pyrrhotite, pentlandite and chalcopyrite.

SOMERSET COUNTY

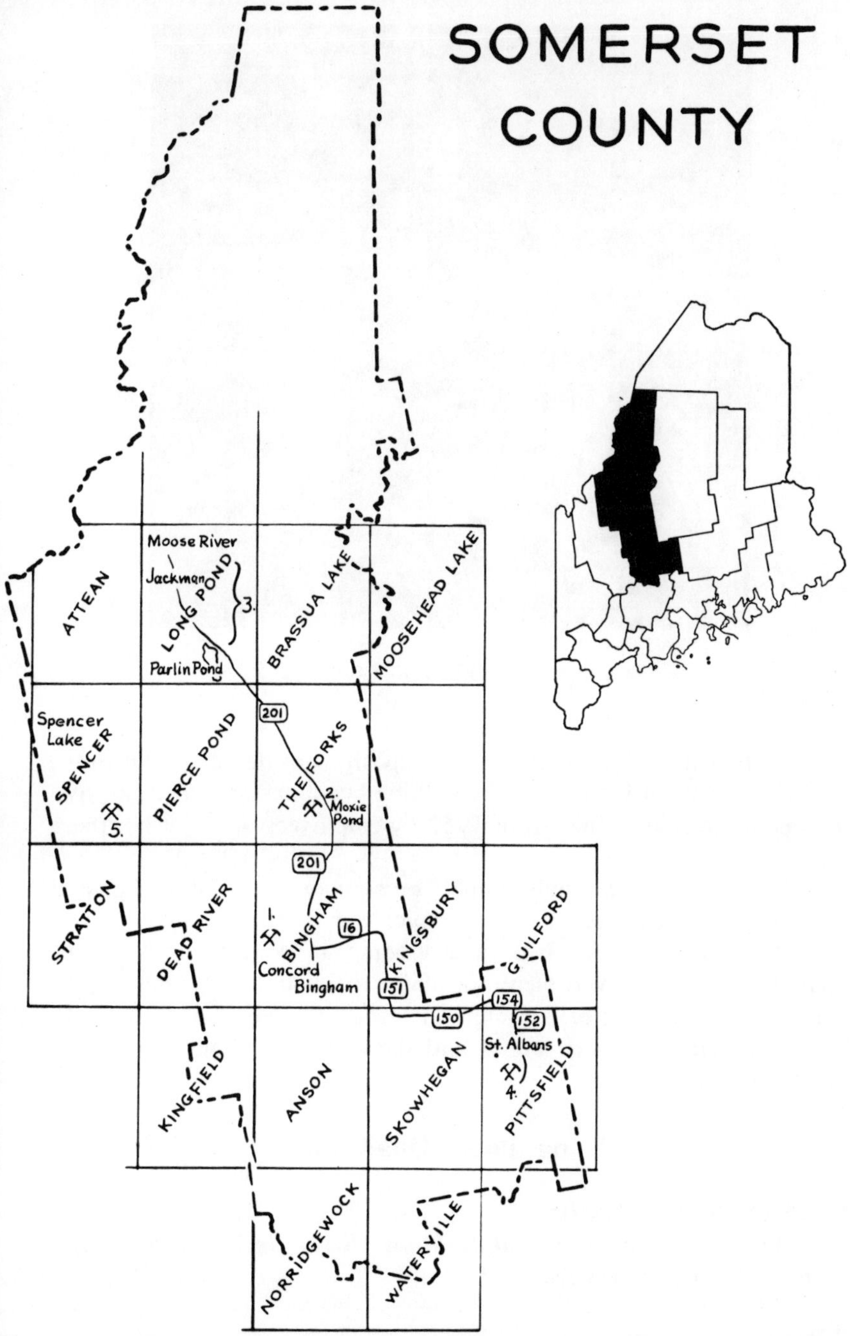

Indian Petroglyph at Embden

My hunch is that this magmatic deposit, considered similar to the nickel prospect at Union, will be making headlines one of these days. It was discovered in the fall of 1952 by employees of Great Northern Paper Co.

When non-professionals stumble on a major find, it should give us rockhounds a sense of mission to spice our treasure hunts.

The head of The U. S. Bureau of Mines in Washington wrote me that the Federal Government considers rockhounds of great importance in the locating of our mineral resources. He said most of the finds, especially of gem stones and the cutting of gems, have been made by amateurs.

Long Pond Quadrangle

JACKMAN TOWNSHIP

Map 3 . . . *Region including Moose River, Jackman and Parlin Pond.* Reported: fossils.

Maine's greatest abundance of fossils extends from here, east across Piscataquis County. They occur along Moose River and on the north flank of Old Squaw Mountain. A highly fossiliferous ledge, containing fully preserved brachiopeds imbedded in grayish or brownish sandstone, occurs at Bean Brook, two miles north of Parlin Pond. Fossiliferous boulders are common all along the woods road which leaves the main highway just north of the pond. Pass around the end of the pond and follow Parlin Stream about 1½ miles to the falls.

Pittsfield Quadrangle

ST. ALBANS TOWNSHIP

Map 4 . . . *Farnum, French, Gould, Indian Lead, New St. Albans, New York and New England, Orient, Porter and St. Albans Quarries.* Reported: galena (lead). Here and there silver, copper and at St. Albans Mine, gold-trace.

Spencer Quadrangle

SPENCER LAKE REGION

Map 5 . . . *Near Eustis on the Arnold Trail.* Reported: large quantities of asbestos near Little Spencer Stream, three miles from the outlet of Spencer Lake. Reported: deposit of asbestos in serpentine. The asbestos is blue material with a greasy luster in the fractures of masses of dense, dark green serpentine. The serpentine shows closely-spaced fracture planes in every conceivable direction.

WALDO COUNTY

Pearls! Near Belfast, as in brooks in many parts of Maine, Basil H. Newell of Belfast has been collecting fresh water pearls for twenty-five years. The colors are pink, white, blue and greenish. He has sold some of his best to Tiffany's, one netting him $120.00. His largest find, two-thirds the size of a marble, was estimated at $500.00. He lost it when jostled on a crowded street in Bangor.

The pearls I get myself from the salt water mussels here in Boothbay Harbor are too small to set. But I have devised a way to use them, and the few flakes of gold I have to show for as many hours of ditch-diggers' labor panning in the Swift River. In each of the tiny treasure chests I used to make to hang on charm bracelets, perched on top of the bits of various gems, was one pearl and one flake of gold. Savin' women, we Mainiacs, of whom I am one by adoption.

WALDO COUNTY

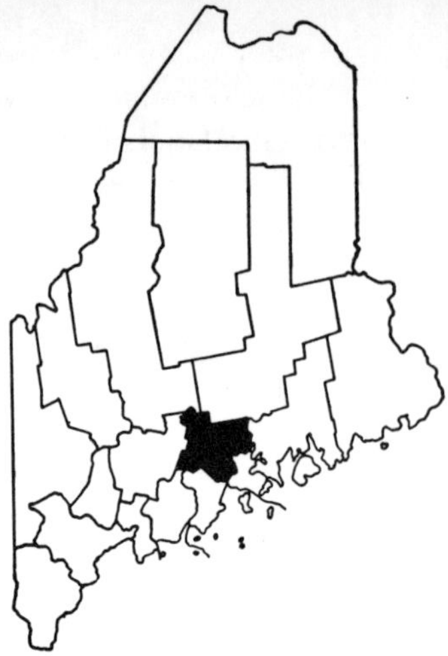

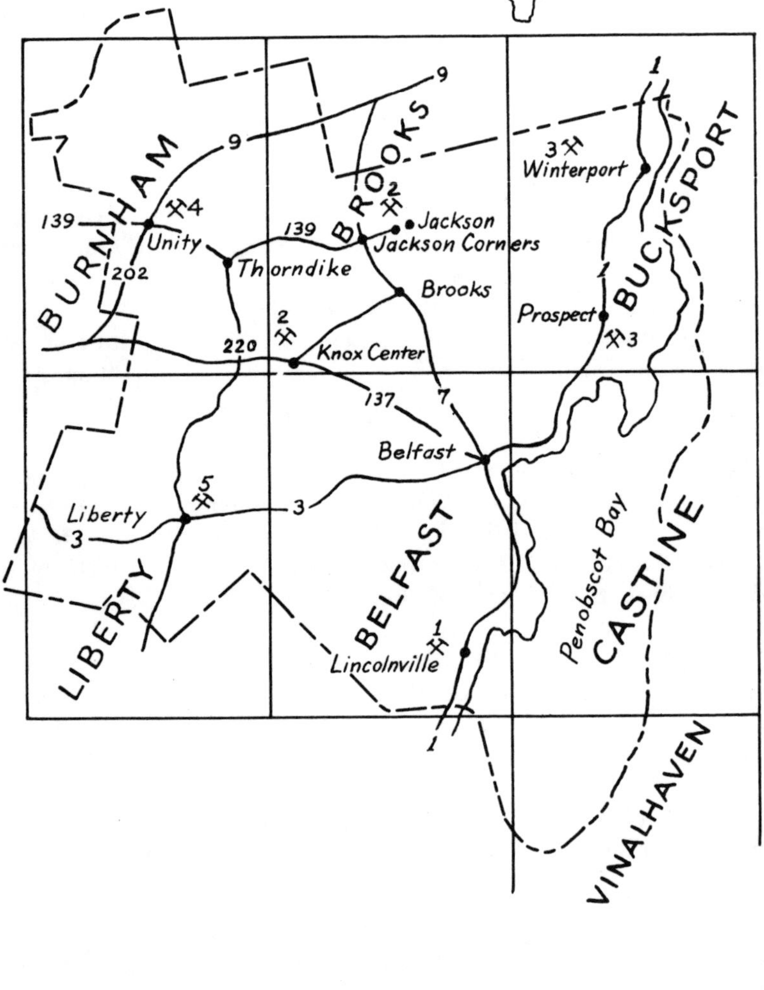

Belfast Quadrangle

LINCOLNVILLE TOWNSHIP

Map 1 . . . *Lincolnville*. Reported: galena (lead), sulphuret of iron (lead, copper and silver). This occurs in a white quartz vein 20-28 feet wide. Shaft 28 feet deep.

Brooks Quadrangle

JACKSON TOWNSHIP

Map 2 . . . *Jackson*. Reported: pyrite, chalcopyrite, sphalerite, galena (gold, silver, copper, lead and zinc). Gold quartz vein 10 feet wide. Slate and schist country rock. Ridge of pyritiferous slate runs immediately north of mine.

KNOX TOWNSHIP

Stone. Reported: silver and gold in a 13-foot open pit on a quartz vein.

Bucksport Quadrangle

PROSPECT TOWNSHIP

Map 3 . . . *Fort Knox*. Reported: copper, lead, silver and gold in a shaft 70 feet deep.

WINTERPORT TOWNSHIP

Waldo.

Burnham Quadrangle

UNITY TOWNSHIP

Map 4 . . . *Unity*.

Liberty Quadrangle

LIBERTY TOWNSHIP

Map 5 . . . *Waldo*.

Three mines, Unity and the two Waldos and none of us knows what is in them! A chance for you to pioneer.

WASHINGTON COUNTY

Here is America's land of the rising sun. Eastport is this country's most easterly point and it is here the morning sun first strikes. This "Down-East" port juts into Passamaquoddy Bay of the twenty-foot tides. We all still hope that some one of these days the Federal Government will harness those gigantic tides to give us cheap power for profitable recovery of Maine's mineral wealth and cheap power for all of the other worthy purposes which would boost Maine's economy and enable many of her youth to remain in this best of all possible lands and still eat.

Write a letter to your Congressmen!

If Quoddy were providing that cheap electricity, we could now be getting aluminum from the clays around Passamaquoddy and Penobscot Bays. Although Maine has no ore bodies of bauxite, the only commercial ore of aluminum, these clays should produce 120 pounds of aluminum per average ton of clay. One acre, ten feet deep, would yield 1800 tons of aluminum.

Near Robbinston a geological phenomenon rears 30 feet high out of the water 50 feet off shore. Called Pulpit Rock, it is an eroded sandstone "stack." Mr. Horace Graham is the sixth generation to own it and his Pulpit Rock Farm.

Calais Quadrangle

Interesting cabinet specimens of crystals of prehnite coated with stilbite crystals are to be found at Calais. These specimens fluoresce.

BAILEYVILLE TOWNSHIP

Map 1 . . . *Baileyville.* Reported: gold visible in white quartz.

Cherryfield Quadrangle

ADDISON TOWNSHIP

Map 2 . . . *Eagle Island.* Reported: ?

CHERRYFIELD TOWNSHIP

Map 3 . . . *Britton.* Reported: galena (lead, copper and silver) in a quartz vein. 20-foot shaft.

Cherryfield. Reported: pyrite, pyrrhotite, chalcopyrite, galena, sphalerite, silver sulphurets, gold, quartz (zinc, lead, silver and gold).

WASHINGTON COUNTY

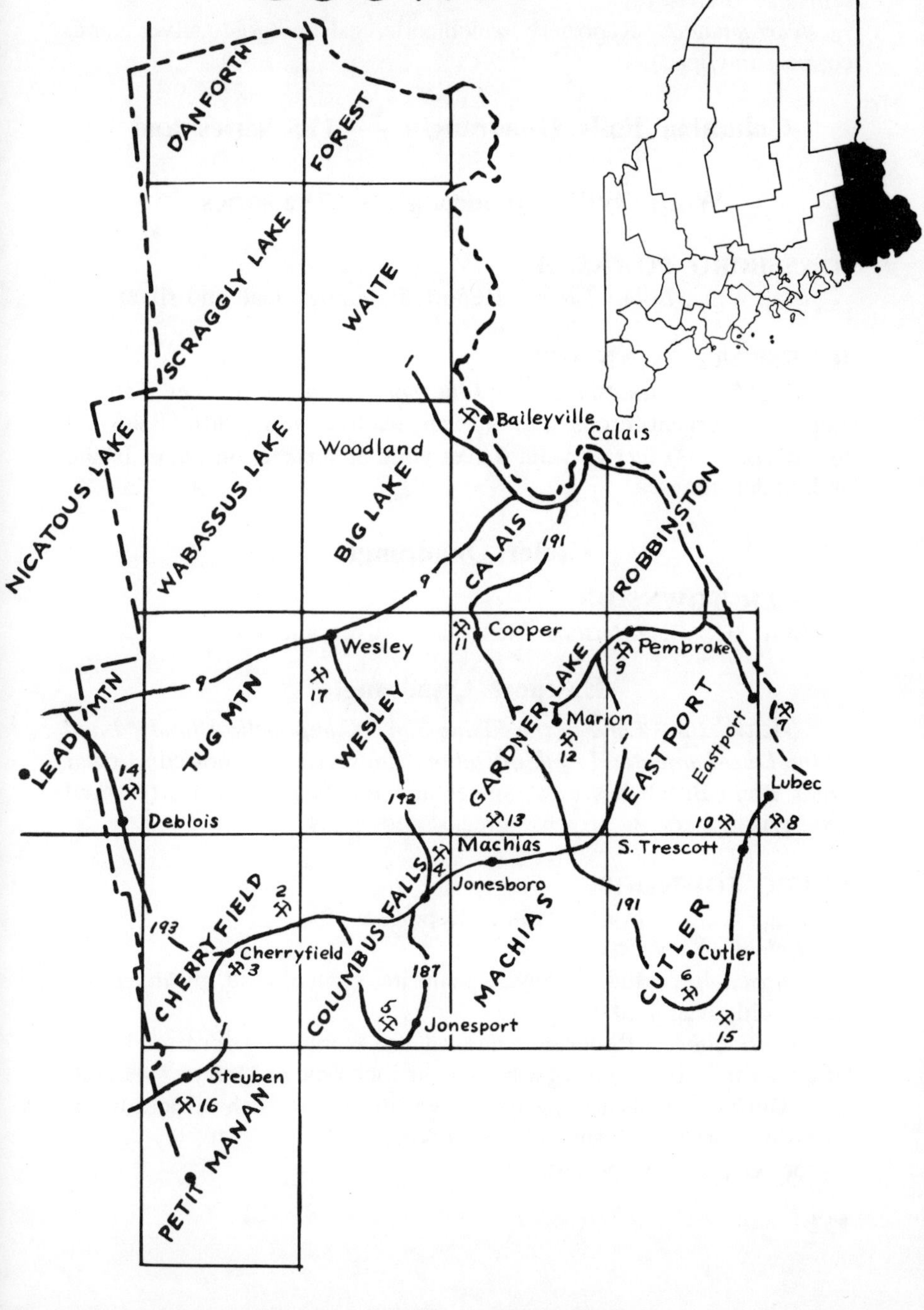

Two compartment shaft 260 feet deep. Vein 12 to 20 inches wide. A fissure deposit with banded structure in country rock of shale and schistose diabase.

Narraguagus. Reported: zincblende, galena (gold, silver, zinc, copper and lead).

Columbia Falls Quadrangle — 115 series and

Whitneyville Quadrangle — 7½ series

JONESBORO TOWNSHIP

Map 4 . . . *Looks Island.* Reported: copper, lead and silver.

JONESPORT TOWNSHIP

Map 5 . . . *Jonesport and Jonesport Extension (Silver Tablet).* Reported: argentiferous galena (copper, iron and gold). Shaft 20 feet deep. 1100 feet of vein, 4 feet wide at surface, on Dunn Island and under the sea.

Cutler Quadrangle

CUTLER TOWNSHIP

Map 6 . . . *Johnson.* Reported: silver and copper.

Eastport Quadrangle

Map 7 . . . *"Land of the Rising Sun," Campobello Island, "Land of the Late-Lamented Franklin Delano Roosevelt."* Reported: Lower Duck Pond Beach has moss agates and fortification agates (marked like prison bars in browns, greens and grays).

LUBEC TOWNSHIP

Map 8 . . . *Denbo Point.* Reported: ?

Lelete. Reported: ?

Lubec. Reported: galena, sphalerite, chalcopyrite, pyrite, limonite (gold, lead and copper).

South Lubec. Reported: natural red paint — a form of hematitic kaolin — oozes out of a bank near the home of Horace I. Hamilton. During hard times the fishermen mix it with fish oil and use it to paint dories and smokehouse doors.

St. George. Reported: ?

PEMBROKE TOWNSHIP

Map 9 . . . *Big Hill Mine.* Reported: galena, sphalerite, chalcopyrite, pyrrhotite, pyrite (lead, zinc, copper, silver and gold-trace). Veins in altered diabase and brecciated basalt; disseminations in country rock; amygdaloid fillings in vesicular basalt.

Pembroke Mine. Reported: azurite, malachite, chalcopyrite, pyrite, galena, sphalerite (silver, lead, copper and zinc).

Big Hill and Pembroke Mines are the basis of my brag that I am part owner of a gold mine in Washington County, Maine! About twenty years ago when Consolidated Dawson Mines of Toronto, Canada, were mining here for copper they brought up a four-foot core of solid, pure gold! The ensuing publicity brought many of us rock enthusiasts running to invest. The hundred shares I bought have never paid a single dividend — and never will. But, it remains my favorite investment.

Map 9 . . . *Straight Bay Mine.* Reported: galena, sphalerite, chalcopyrite, pyrrhotite, pyrite (lead, zinc, copper, silver and gold-trace).

Young's Point. Reported: argentiferous galena (lead, silver and copper). Big Hill and Straight Bay mines are the same vein of zinc, lead, copper and silver which extends under the Cobscook tide and comes to the surface in the Pembroke and Perry basins. This is part of the Penobscot Mining Company's operation which starts at Cape Rosier in Hancock County. Some of the most valuable fossil deposits in the western hemisphere are found here as well as low grade coal.

TRESCOTT-LUBEC TOWNSHIP

Map 10 . . . *Trescott.* Reported: galena (silver, lead, copper and gold-trace).

Gardner Lake Quadrangle

COPPER TOWNSHIP

Map 11 . . . *Calais Molybdenite and Cooper Molybdenite.* Owned by American Molybdenum. Reported: molybdenite, native bismuth, iron molybdate (molybdenum). Molybdenite ("white gold") mineralization follows a north-south joint set in granite, genetically related to granite pegmatites parallel to this set. Segregations of molybdenite up to 10 or 12 pounds have been reported. Some plates up to 2 inches in diameter. Shaft 50 feet deep is trenched by a ditch 225 feet long and 4 feet deep.

Don't let the above overwhelm those of you who are new to this field (I don't worry about the geology professors among my readers!). Commit it to memory. The jargon may open unexpected doors. I know!!

MARION TOWNSHIP

Map 12 . . . *Gardner*. Reported: ?

Marion. Reported: copper.

MARSHFIELD TOWNSHIP

Map 13 . . . *Bowker*. Reported: gold in quartz vein.

Crocker. Reported: lead and silver.

Machias. Reported: silver.

Lead Mountain Quadrangle

DEBLOIS TOWNSHIP

Map 14 . . . *Leighton*. Reported: vein with silver runs easterly, crossing Pleasant River at Saco Falls, thence toward Whitneyville.

Machias Quadrangle

CUTLER TOWNSHIP

Map 15 . . . *Cross Island (Neptune)*. Reported: copper, fluorescent calcite.

Roque Bluffs. Reported: jasper on the beaches here and towards Jonesport. There is a Jasper Beach where jasper abounds. But, it was named after a Captain Jasper!

Petit Manan Quadrangle

STEUBEN TOWNSHIP

Map 16 . . . *Petit Manan and Pillsbury Stevens*. Reported: NOTHING! A small shaft at each location.

These? locales should be a challenge to you, especially as an extra dividend might await you. This, the most indented coastline in the world was known and used by the pirates of the Spanish Main.

Dixey Bull, Captain Kidd, Teach (Blackbeard) and Sam Bellamy, the "Robinhood of the Seas," who established his own empire at Machias, all blew with the trade winds from Cuba, Jamaica or the Bahamas in the West Indies whenever the barnacles and weed growth on the hulls of their sloops and galleons slowed up their nefarious chase.

Here, at the mouth of the Penobscot River or in the sheltered coves of the myriad of islands along the Maine coast, they careened their vessels and scraped the bottoms to glass smoothness. Then they hoisted their Jolly Rogers and blew back again to the Caribbean Sea.

But, not before they had buried their iron pots or brass-bound chests full of Spanish pieces of eight and gold doubloons. If you find one of these grand-daddies of the modern safe deposit box, and every so often someone does, the dividend will be greater than the mineral treasure you set out to seek.

Wesley Quadrangle

WESLEY TOWNSHIP

Map 17 . . . *Bacon and Snow*. Reported: sphalerite, pyrite, chalcopyrite (gold, silver, zinc and copper).

Edwards and Walker. Reported: gold.

Rollins. Reported: galena, sphalerite, pyrrhotite, bornite, pyrite, cuprite (lead, zinc, gold and silver).

YORK COUNTY

Berwick Quadrangle

ACTON TOWNSHIP

Map 1 . . . *Acton Consolidated: Acton, Androscoggin Acton, Boston Acton, Bullion Casco Acton, Deering Acton, Forest Acton and Portland Acton mines*. Owned by Boston Acton Silver Mining Co. Reported: pyrrhotite, galena, pyrite, arsenopyrite, chalcopyrite, sphalerite (lead, silver, gold and copper).

Map 2 . . . *Dirigo Silver*. Reported: silver and lead. Shaft 55 feet deep.

Eldorado. Reported: silver.

Lebanon Acton. Reported: galena, native silver, pyrite (lead and silver), soapstone. Three compartment shaft 45 feet deep.

Mineral Hill and Morris and Silver Wave. Reported: copper. Three shafts and an addit. No. 1 118 feet deep.

LEBANON TOWNSHIP

Map 3 . . . *Maine Acton and York County*. Reported: lead in quartz stringers with ore encountered at 19 feet depth.

YORK COUNTY

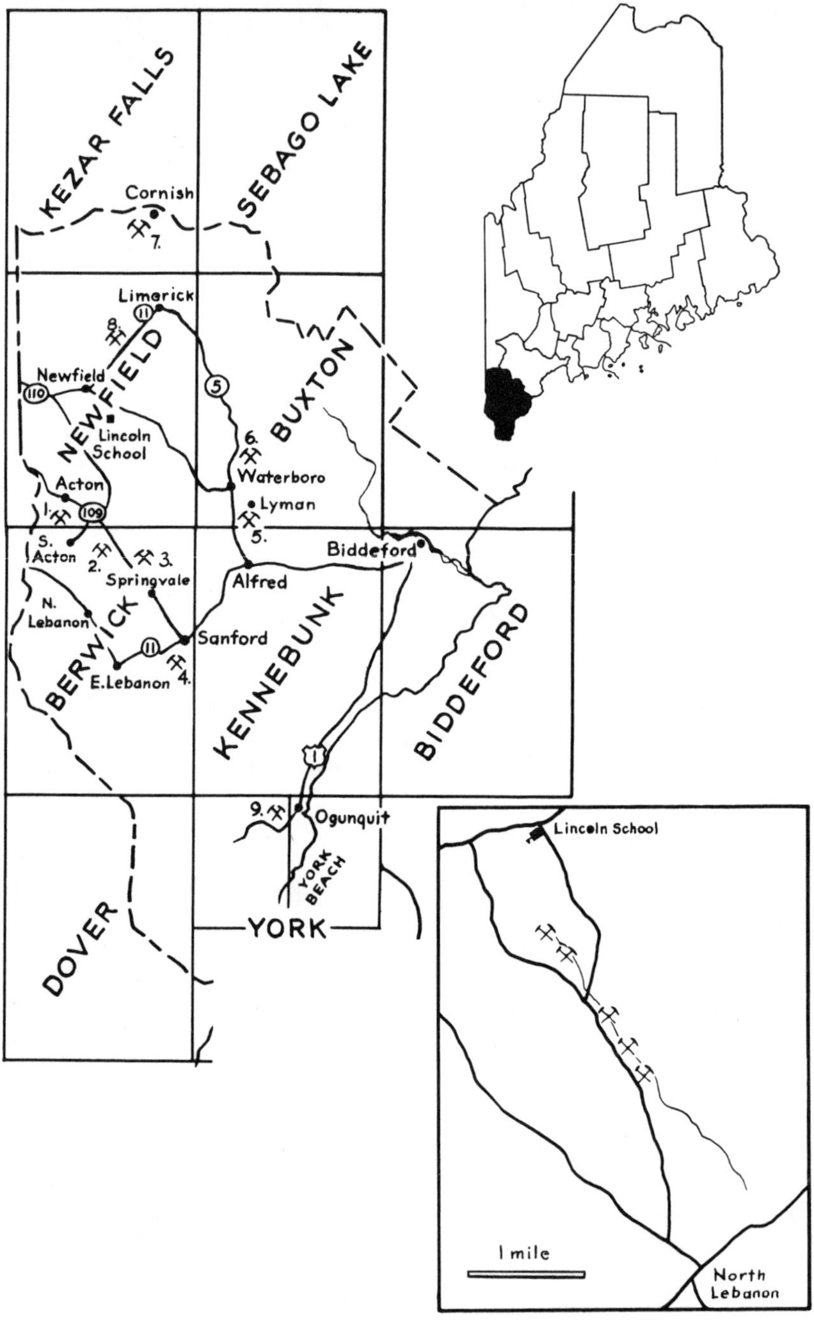
KEZAR FALLS
SEBAGO LAKE
Cornish
7.
Limerick
11
8.
NEWFIELD
Newfield
110
Lincoln School
5
BUXTON
6.
Waterboro
Lyman
Acton
1.
109
5.
S. Acton
2.
3.
Springvale
Alfred
Biddeford
N. Lebanon
BERWICK
11
Sanford
E. Lebanon
4.
KENNEBUNK
BIDDEFORD
1
9.
Ogunquit
YORK BEACH
YORK
DOVER
Lincoln School
1 mile
North Lebanon

SANFORD TOWNSHIP

Map 4 . . . *South Sanford.* Reported: scheelite, molybdenite, vesuvianite, molybdite, powellite (tungsten and molybdenum).

This is the general location of the world-famous deposit of vesuvianite crystals (idocrase). It is mentioned in the Dana-Hurlburt Manual of Mineralogy (geologists' bible).

All important museums, public or private, all over the world have specimens of Sanford, Maine vesuvianite.

As the name implies, this mineral was found in ancient times near the crater of Mt. Vesuvius in Italy. The specimen I obtained at the smoking crater now rests next to the one I found at Sanford.

The gemmy green variety from the far west is called californite. The Sanford material is brown, so is the Italian, and crystallizes in four-sided prisms with pyramidal terminations. It is vitreous, 5.6 hardness and is subtransparent to translucent.

In this area an exotic mineral assemblage grades into a banded lime-silicate gneiss (there's a "nice" word to impress with!). The vesuvianite deposit seems to be sandwiched between granite.

Buxton Quadrangle

LYMAN TOWNSHIP

Map 5 . . . *Brock Prospect.* Owned by Henry Brock, Alfred. Reported: beryl, feldspar and muscovite.

WATERBORO TOWNSHIP

Map 6 . . . *Caton.* Owned by Walter Caton, Saco. Reported: feldspar and quartz.

Newfield Quadrangle

CORNISH TOWNSHIP

Map 7 . . . *Berry.* Reported: scheelite (tungsten), calcite, grossularite, scapolite, idocrase, pectolite, diopside. Bedrock is lime-silicate gneiss and quartz-biotite-feldspar schist and gneiss (yes, it's pronounced "nice"). Several test pits have revealed minor amounts of scheelite (tungsten) in skarn zones. The largest masses are up to 2 inches across.

Pease Mountain. Reported: scheelite, diopside, grossularite (garnet), idocrase (tungsten). Traces of scheelite in skarn of lime-silicate gneiss at contact with quartz syenite pegmatite.

NEWFIELD TOWNSHIP

Map 8 . . . *Bergandahl*. Reported: molybdenite, chalcopyrite, sphalerite, pyrrhotite, pyrite and pentlandite. Small cut made in the hillside.

Newfield Tin. Reported: tin.

Piper Hill. Reported: scheelite (tungsten) and grossularite (garnet).

York Beach Quadrangle

Ogunquit. Reported: gold. Robert Doyle, our State of Maine Geologist, says that the report of gold here is based upon a very small amount being found in a brook.

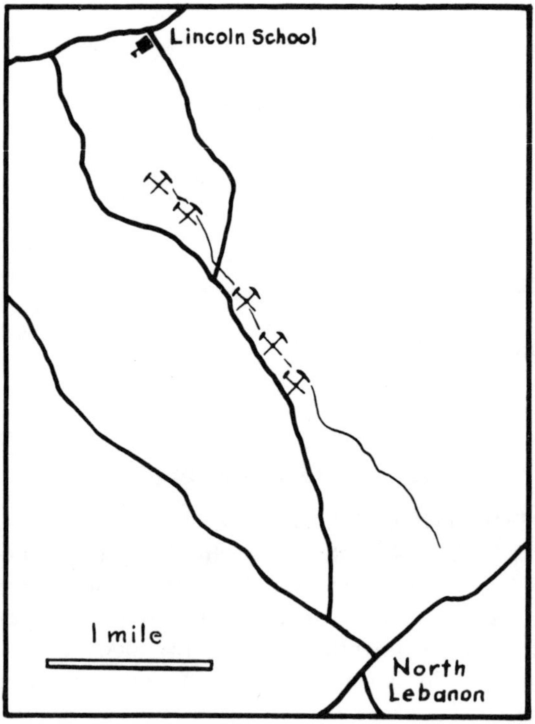

6. Back With Your Treasures

Empty the bags! Rediscover the fine examples you collected with so much hope and gladly toted down the mountain. Where are they, those excellent cabinet specimens? You know there were some gemmy pieces! You proved it by the lick test (an eager lap of your wet tongue). The one you found in the brook glowed with color (when in the brook). Where?

Oh! Here is that quartz crystal, all six sides complete, terminating in a neat pyramid. There is the huge chunk of rose quartz with several "windows." At least four of the transparent, dreamy pink nodules will cut into gemstones. One small section of that massive beryl crystal is unquestionably gem aquamarine. Oh, what a beautiful green — these slender little tourmaline crystals! Those aren't flaws! They are "feathers."

The rest? Throw them away? Fill that hole in the driveway? Never! Here are some of the many possible uses for your hard won treasures, from easiest to hardest:

DECORATIVE

Rock gardens.

Flower arrangements.

Lamps. Select a large striking specimen (imagine the light reflecting on mica or gold veins or through rose quartz or beryl!) and cement it on to an oak block, rubbed with a touch of dull color. Attach to the wood whatever fixtures you will need to accommodate the electricity and shade. Choose a neutral shade large enough to provide a pleasing proportion for the whole lamp.

Bookends, doorstops and paper weights, either in the rough or polished. If rough pick a specimen which stands on a firm base. Cement three small felt circles to the bottom to protect the table. If polished, either have a lapidary prepare it for you or learn to do it yourself. I have described the lapidary art briefly below. There is quite a vogue now for mineral paper weights. When cut into squares or hemispheres they command from $45.00 to $100.00 each.

Crystal-gazing balls. Become a fortune teller! I crave a large one of rose quartz and am always on the lookout for a very pink, large and limpid piece which is perfectly transparent and flawless. No clouds in my future if I can help it! Asteriated, too. I want a star in my future! You can buy sphere cutting equipment.

MINERAL COLLECTIONS

You can make this hobby as simple and inexpensive or as complicated and costly as you want to. Creep, then walk, before you run.

If you write to Ward's Natural Science Establishment, Box 24, Beechwood Station, Rochester, N. Y., they will send you their catalog with illustrations of equipment they sell for preparing and displaying a mineral collection. And, if your treasure hunts have been too lean, you can buy mineral specimens from them.

Here are a few rules whatever the type and magnitude of your collection:

1) The standard size for non-crystalline (massive) specimens is 3" x 4" on top and 1" to 2" thick.

2) Crystals will have to vary in size. Choose for perfection of form, color, examples of cleavage, twinning, etc.

3) Appearance of specimens should be clear, clean and shapely. Remove limonite (iron rust) stains with a saturated solution of oxalic acid, one cup to a gallon of water in a glass or crockery (not metal) container. It is poison, so don't get it on or in you.

4) Label every specimen — at least with its name and exact locality from which it was collected. The more information the merrier — composition, structural features, associations and classification. Use the Dana numbering system as described in any one of his excellent books, for instance "Textbook of Mineralogy." First give the Dana number, then your own. For example, Pyrite. Your first specimen, 2911-1, your sixth, 2911-6. Paint a neat white rectangle on an inconspicuous place on the back of the specimen. When dry, write the number on it in black India ink. Varnish. Keep a catalog of your numbered specimens with the information recommended above.

5) Have a number on the label corresponding to one on the-specimen.

Various Types of Collections

Greatest variety; — from Maine, — from all over the U. S., — from all over the world; varieties from each geological location; — in Maine, — in the U. S., — in the world; examples from specific locales; — in Maine, — in the U. S., — in the world; crystals only;

widest range of color; microspecimens; fluorescent specimens with a dark cabinet and ultra-violet light to show them; fine cabinet specimens; rarities and oddballs.

QUARTZ ODDITIES

Asteriated (starred); crystal clear; with blue or green tourmaline; crystals included; tabular; rutilated (with hair-like inclusions of rutile); parallel grouped; peduncles; sceptered; green, amber and gemmy, smoky crystals; phantoms (one or more crystals inside a larger one. It is well to slab and polish front and back to show inside).

TOURMALINE ODDITIES

Rare coloring; bent crystals; hollow crystals (some can be strung like bugle beads); red-tipped crystals; fan-shaped crystals; flattened crystals; zoned color sections; nice terminations; showing coating of pocket cookeite.

JEWELRY

For your choicest finds, cut them into gemstones. Set them in rings, bracelets, cufflinks, or bola ties. Wear them with ill-concealed pride. Or sell them.

The Easy Way

Nothing easy about it, just easier, is to tumble-polish the stones then cement them into bought jewelry findings. "Findings" — another word of the jargon for you to learn — the ring blanks, earring backs, pin stems and stone cappings.

The University of Maine Extension has instructors in the field; Kay Baird, for one, to teach this. You might write them for a bulletin on it.

Most of the mineral stores in Maine I have listed in Chapter 4 will have the various necessaries for tumbling and for this method of setting.

STEPS IN TUMBLING

1) Break up your large material with a mallet.

2) Select pieces nearly as small as the stones you want and of the same relative hardness.

3) Throw them all into an electrically rotated barrel, the speed of which you can regulate.

4) Throw in after them the coarse grit, silicon carbide No. 80.

5) Check every few days to see if ready for a change of grit. Wash between changes of grit.

6) Now, corundum powder, fine, No. 400, after, perhaps, two to three weeks.

7) The whole process takes six to seven weeks, depending on the hardness of the material.

8) Increase the speed as indicated.

9) The last couple of weeks are for the final polishing with cerium or tin oxide.

10) Remove, scrub, admire and select to set into jewelry.

MAKING THE JEWELRY

This type is just a matter of gluing the stones to the findings. For instance, bracelets:

1) Fit each stone into a cap. Cement.

2) Attach each capped stone to a link in a bracelet chain with a jump ring, using pliers.

3) Wear.

YANKEE INGENUITY

Inventive, like so many of our Maine people, the late and sorely missed Stearns Bryant did not send for his tumbling outfit from a catalogue. Rube Goldbergish, the one he made himself, but it was highly efficient. My drawing shows:

A converted Model T Ford rear end, which was run by an electric motor (not shown).

A ¾-inch plywood turntable.

A vertical frame with a horizontal rod attached to the frame above the turntable.

A fat airplane tire, big enough to hold plenty of gem material, suspended on the rod. By moving the tire nearer the center for slower or to the outside for faster, he could regulate the speed. Like "snap the whip."

Perched on top of the rod, a can of water with a tube running from it to drip dust-laying water.

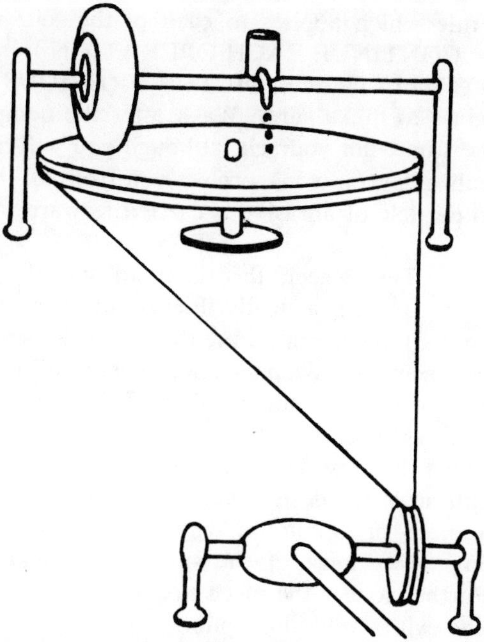

STEARNS BRYANT'S GEM TUMBLER

JEWELRY

The Hard Way

If anything I may say in this sketchy description of the ancient lapidary art or the high skill of making fine handmade jewelry gives you the imporession it is simple — perish the thought!

Every step of the process, and hundreds are required, is the hardest, the most onerous, the most skill-requiring, the most pains-taking and the most delightful pursuit known to man.

CUTTING A GEMSTONE

I'm going to explain the four steps in cutting a cabachon. Cabachon is from the French *caboche,* a precious stone of convex or rounded form which has been polished, but not cut into facets.

I'll skip the difficult art of facetting, assuming that by the time you have mastered the cutting of cabs you will have picked up more information about diamond or emerald cuts than I can hope to give you here.

A basic rule which applies to each of the four major lapidary phases is this: CONTINUE EACH OPERATION LONG ENOUGH TO REMOVE ALL TRACE OF THE PREVIOUS STEP. Use a magnifying glass to make sure. Wash after the completion of each operation (the stone, not yourself, although you will need to because your heavy rubber apron won't protect you from all the lovely mess) to be sure no particle of abrasive is carried forward to the next step.

Steps:

1. Sawing. This wheel, the diamond saw, has diamonds all around its edge. It needs a steady flow of lubricant, two parts kerosene and one part motor oil, into the cut you are sawing.

   a) By means of sawing, section a slab ⅛" to ¼" thick in a way to take advantage of the best color or pattern your rock will yield.
   b) Select the most desirable portion of this slab and mark the outline of the desired cabachon, freeform or regular, with a piece of aluminum wire.
   c) Saw away most of the surrounding material.
      Further reduce the piece to conform with the desired outline with "nibbling" pliers.

You now have a preform. Fun to use the jargon, isn't it?

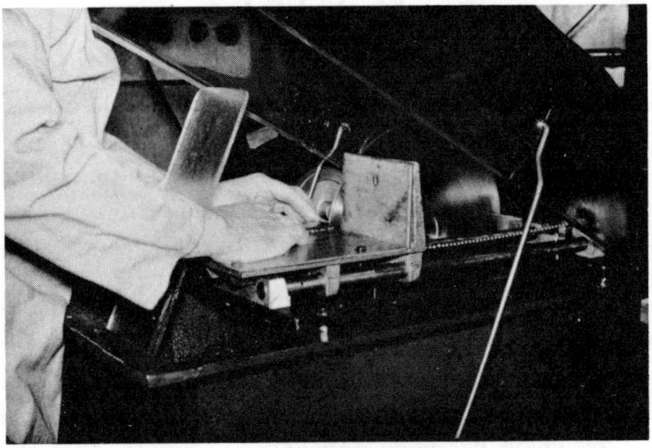

Diamond Sawing

2. Grinding, with a steady flow of water to eliminate dust and the generation of too much heat, which might fracture the stone. Replace the diamond saw with the coarse grinding wheel.

a) Hold the stone with the tips of your fingers and rotate constantly against the edge of the revolving wheel.
b) Inspect frequently.
c) Bring your cabachon gem as close as possible to the final form during the rough grinding operation.
d) Do not try to hurry the stone — because you can't.

Replace the coarse grinding wheel with the fine one. Continue as before until any minor imperfections of form or marks are removed and the cab has taken its final form. The major ones should have disappeared with the coarse grinding.

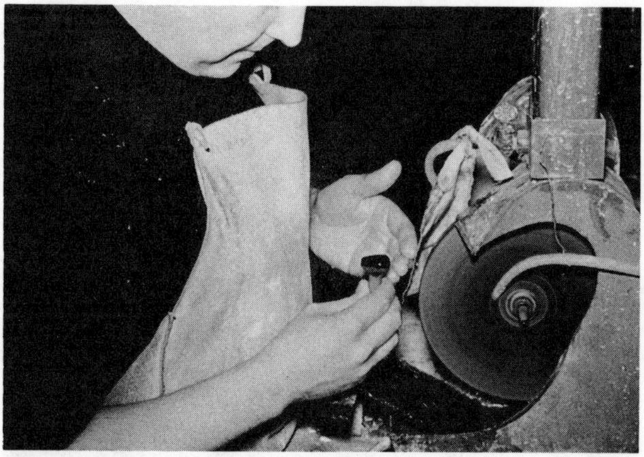

Fine Grinding

3. Sanding, done dry. Replace the fine grinder with the sanding wheel.

a) Cement your cab onto a "dop stick" for easier handling when sanding and polishing.
b) Move the stone around with a sweeping motion on the lap (surface of the sanding wheel) and lift frequently since this process develops considerable heat.
c) Continue the sanding until all ridges and grinding marks are gone, and just a little bit longer for good measure.

4. Buffing. Replace the sanding lap with the felt buffing lap, well wet down with water.

a) Dissolve a spoonful of polishing agent in water to the consistency of thick cream and apply to the moistened felt buff.
b) Hold the stone against the revolving felt with considerable pressure and move about to bring the entire surface of the gem in contact with the polishing agent.
c) After you are sure the stone has the best possible polish, remove it from the dop stick by inserting the point of a pocket knife between the stone and the stick and applying pressure.
d) Scrape off any dopping wax with the knife. If necessary, remove any traces of the dopping cement with wood alcohol or carbon tetrachloride.

NOW — set your masterpiece!

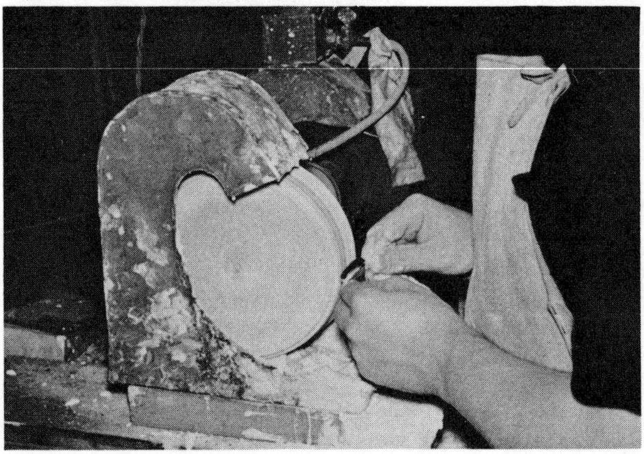

Polishing

MAKING A PIN-SEASHAPES

To see how it's done, let's follow the steps in making a pin-motif Seashapes with two freeform cut green tourmalines prong-set in 14K gold.

Tools:

a) Jeweler's saw with spare saw blades.
b) Files of various shapes and fineness.

c) Bench vise and hand vise.
d) An assortment of pliers.
e) Jewelers' hammers.
f) Drill.
g) Copper tweezers, copper tongs and copper pickling pan.
h) Prest-O-Lite soldering torch with adjustable tip and cylinder of gas.
i) Charcoal soldering block.

Materials:

a) Sheet of medium-gauge 14K Hamilton yellow gold from Handy and Harmon, N. Y.
b) Pin assemblage: joint, cotter pin, safety catch and pin stem.
c) Two polished green tourmaline gemstones.
d) Handy and Harmon flux and solder for 14k yellow gold.
e) Pickling solution: I part sulphuric acid to 10 parts water. *Add the acid to the water.*
f) Bowl of rinsing water.

How to:

a) Draw on paper the design for the pin and the prongs to hold the stones (note that two of them are shaped and sawn out to suggest breaking waves).
b) Scribe (draw) on the sheet of gold, indicating the outlines of the holes to be sawn out under the stones to let the light up through.
c) Clamp the sheet of gold in the vise.
d) Saw out the pin shape and the prongs. Drill holes to accommodate the saw wherever there are to be holes in the surface of the pin. Saw out the holes.
e) File all edges smooth.
f) Accurately measure the stones to judge where to place the prongs for a firm setting.
g) Solder the prongs in place. On the back, solder the pin assemblage. Choose a pin stem long enough to extend the width of the pin but not beyond the edge. Solder the joint and catch above the center line for good balance.

How to hard solder:

a) Because solder will not bridge a gap, ensure a snug fit by cleaning the surface and the object to be soldered. Do this by scraping, filing or pickling. After each soldering operation grasp the piece with the copper tongs and plunge into the copper pan of the hot pickling solution. Rinse.

b) Apply flux to the spot on the surface of the pin and to the object to be soldered on to it. Not too much.

c) Place your work on the soldering block.

d) Dim the light so in the semi-darkness you can see the color of the metal and thus estimate the needed degree of heat.

e) Apply heat from the torch GRADUALLY until the water in the flux has evaporated.

f) Keep a soft blue flame in motion all the time and off the joint as much as possible.

Soldering

Polishing Metal

g) When the joint shows dull red, concentrate the heat upon it until, and only until, the solder flows.

h) When all parts are soldered to the pin, clean in the pickle, rinse and let dry, then file away rough spots and smooth with fine abrasive paper. Polish before setting the stones.

How to polish:

a) Use a separate buff on your arbor, or electric motor, for each polishing compound used (and for each different metal).

b) Eliminate deep scratches with tripoli on a muslin buff.

c) Get your final polish with rouge or white diamond compound on a muslin buff.

Setting the stones and securing the pin:

a) Place the stones where planned and firm the prongs against them with a burnisher.

b) Slip the non-pointed end of the pin stem into the slit in the joint. Hammer the cotter pins through the holes in the joint. File Smooth.

Your pin is finished! After instruction (the above is too sketchy to serve) and about six years of practice you will be able to make this pin, expertly enough, so it will sell for a high price.

CONCEIVING A DESIGN

You wouldn't want to make jewelry that looks like that of everyone else; so, let your imagination soar. Years ago when I was designing my "Jewelry In The Maine Mood," I used to let nature, as I saw it along the Maine coast, suggest ideas.

Here are pictures of waves, tide treasures, and seashapes, and the jewelry they inspired.

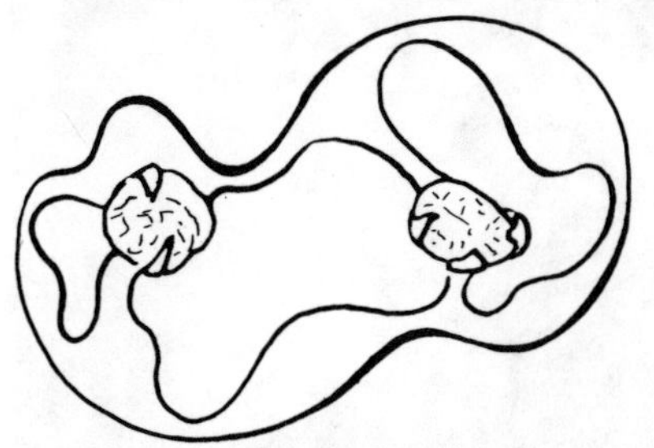

Pin Motif – Seashapes

Inlet of the Sea, Boothbay Harbor

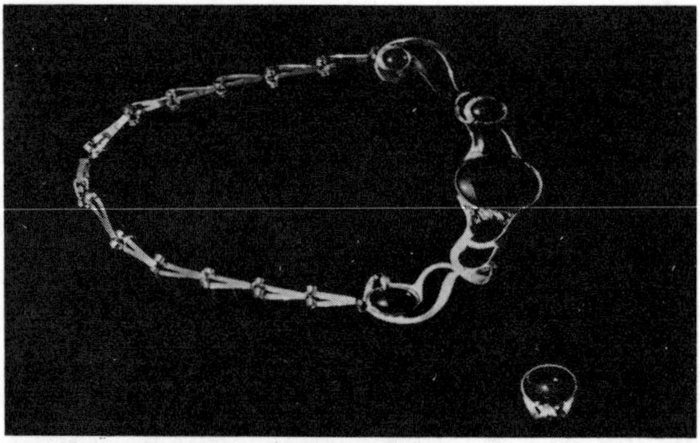

Necklace and ring-motif Wave

Waves on Georgetown Island

Tide Treasures arranged on the dock
of the author's island

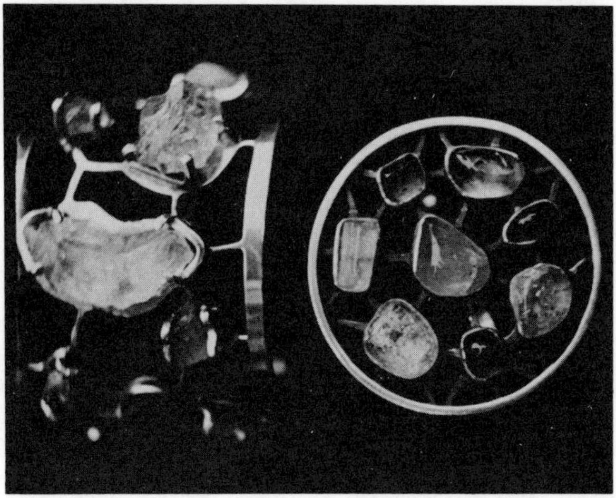

Pin and bracelet-motif Tide Treasures

TRAIL'S END

Hasn't it been fun, our treasure hunt throughout Maine from START: learning that you do walk on jewels, to FINISH: how to use your treasure finds in various ways? We all but looked under every stone in the State, didn't we?

Index